AF360886

Green Coal Mining Techniques 2020

Green Coal Mining Techniques 2020

Editor

A.J.S. (Sam) Spearing

MDPI • Basel • Beijing • Wuhan • Barcelona • Belgrade • Manchester • Tokyo • Cluj • Tianjin

Editor
A.J.S. (Sam) Spearing
School of Mines,
China University of Mining & Technology,
Xuzhou, China

Editorial Office
MDPI
St. Alban-Anlage 66
4052 Basel, Switzerland

This is a reprint of articles from the Special Issue published online in the open access journal *Energies* (ISSN 1996-1073) (available at: https://www.mdpi.com/journal/energies/special_issues/CM_2020).

For citation purposes, cite each article independently as indicated on the article page online and as indicated below:

LastName, A.A.; LastName, B.B.; LastName, C.C. Article Title. *Journal Name* **Year**, *Article Number*, Page Range.

ISBN 978-3-03943-589-0 (Hbk)
ISBN 978-3-03943-590-6 (PDF)

Contents

About the Editor

A.J.S. (Sam) Spearing works in the School of Mines at China University of Mining & Technology and is an Adjunct Professor at the Western Australian School of Mines, Curtin University, where he was the Director. He has extensive underground mining design and support experience and focuses on risk and safety. He has co-invented seven products that are routinely used in mines and has published over 50 peer-reviewed papers in addition to several books and book chapters. He has both industry and academic experience.

Preface to "Green Coal Mining Techniques 2020"

Globally, coal mining is slowly reducing as alternative renewable energy sources such as solar energy become more cost effective. Coal will, however, still be mined for years to come, especially the higher grade bituminous coal that is needed for steel production. It is therefore important that research into green and sustainable coal mining continues so that safety continues to improve and the environmental impact of mining is reduced or eliminated. Collaborative research is important and is growing, especially between universities both domestically and globally. Green mining, especially environmental protection and the social license to operate, have become the front and center of the apparatus to reduce the impact of mining on the environment, especially since the UN published its Sustainable Development Goals. Although some will say that not enough has been done, the progress over the last decade has been significant and progress continues. Mining is key to our survival along with water, food and energy, and the fourth industrial revolution has the potential to benefit the mining industry the most. It is an exciting time to be in mining.

A.J.S. (Sam) Spearing
Editor

Article

Sustainable Rehabilitation of Surface Coal Mining Areas: The Case of Greek Lignite Mines

Francis Pavloudakis [1], Christos Roumpos [1,*], Evangelos Karlopoulos [2] and Nikolaos Koukouzas [2]

1. Mining Engineering Department, Public Power Corporation, 104 32 Athens, Greece; f.pavloudakis@dei.com.gr
2. Research Department, Chemical Process & Energy Resources Institute, Centre for Research & Technology Hellas, 15125 Athens, Greece; karlopoulos@lignite.gr (E.K.); koukouzas@certh.gr (N.K.)
* Correspondence: c.roumpos@dei.com.gr; Tel.: +30-697-979-9291

Received: 2 July 2020; Accepted: 31 July 2020; Published: 3 August 2020

Abstract: Surface lignite mines of the Balkan Peninsula face growing pressure due to the CO_2 emissions reduction initiatives, rapidly increasing renewable-power capacity, and cheap natural gas. In this frame, the development of a modern mine land rehabilitation strategy is considered as a prerequisite for mitigating the social and economic impacts for the local communities. In the case of western Macedonia lignite mines, these prospects are investigated based on a PEST (political, economic, social, technological) analysis of seven alternative land uses. Urban (industrial) development, green houses, and industrial heritage parks are considered as the most appropriate land uses for mitigating the socioeconomic impacts due to the loss of employments. For the land uses occupying large areas (i.e., agriculture, forestry, livestock farming, and photovoltaic parks), an optimisation algorithm is proposed for determining the mix of land uses that maximise revenue, equity, and natural conservation and minimise investment. The algorithm was applied using the opinions of 10 experts, who were involved in mine land reclamation projects carried out in the western Macedonia region in the recent past. According to the results obtained, photovoltaic parks are rated as a more attractive investment than extensive agriculture, as regards the anticipation of revenues, while livestock farming and forests are necessary to safeguard the ecosystem's functions.

Keywords: coal; lignite; mines; reclamation; rehabilitation; land use; decision-making; optimisation; sustainability

---

1. Introduction

The European Green Deal is the response of the European Union (EU) to the climate and environmental-related challenges that are the defining task of this generation. It is a new development strategy that aims to transform the EU into a fair and prosperous society with a strong and competitive economy, where growth is decoupled from intensive resource use and there are no net emissions of greenhouse gases [1]. By signing the Paris Agreement, the EU made an official commitment to undertake certain measures for limiting global warming to "well below 2 °C" [1–3]. Then, in November 2018, the European Commission released a long-term strategic vision to reduce greenhouse gas emissions, between 80% and 100% compared to 1990. In fact, the ambitious goal of a 100% reduction of greenhouse gases represents a climate-neutral economy [4].

In order to achieve this goal, the EU will need to rapidly decarbonise its power sector. However, the implementation of this strategy poses significant technological, economic, and social challenges [4]. Coal infrastructures exist in 108 European regions. It is estimated that the coal sector currently employs about 237,000 people (185,000 work in coal/lignite mines). Poland employs about half of the workforce of coal/lignite mines and thermal power plants, followed by Germany, the Czech Republic, Romania, Bulgaria, Greece, and Spain [2].

Coal and lignite-fired thermal power plants face overwhelming regulatory pressures to lower greenhouse gas emissions. Power plants also face increasing competition from growing renewable-power capacity and from lower natural gas prices, trends that are developing against a backdrop of stagnating demand. While European year-ahead (2019) coal prices have dropped to about $66 per ton from more than $100 a year ago, EU carbon permits have escalated fivefold since 2017. As a consequence, the current cost of burning coal is significantly higher [5] and coal-fired power plants are becoming unprofitable as they have experienced losses worth €6.6 billion in 2019, according to a new study by the British think tank "Carbon Tracker" [6]. The most exposed utilities are the German RWE, the Czech EPH, and the Greek Public Power Corporation SA (PPC) who could haemorrhage about €2 billion in 2019 [2,7].

The above-described situation is particularly true for the lignite-intensive regions of the Balkan countries that have to be prepared for the reduction or phasing-out of mining and lignite power generation activities [7]. The dependency of these regions on the lignite exploitation resulted in limited growth of other economic sectors. This is related to a socioeconomic phenomenon known as "lock-in": A tendency for incumbent industries to actively resist the diversification of the local economy in order to avoid competition for economic resources [8]. This fact, combined with indirect impacts of mining on the quality of environment, properties' value, and health, leads to a vicious cycle, whereby the deteriorating attractiveness of the regions for new investments makes them ever more reliant on the lignite exploitation [8]. However, many coal dependent regions perceive decarbonisation driven by environmental considerations only as a process with adverse socioeconomic impacts, which expose local communities to unnecessary turbulence with unpredictable flow-on effects [7].

Despite the aforementioned obstacles, the reduction or phasing-out of mining and coal power generation activities is considered as one of the most cost-effective methods to achieve emissions reduction due to both market-driven and regulatory trends [6]. Under these circumstances, the economic importance of coal mining has decreased. Coal mining regions have lost a considerable part of their economic base and are facing structural changes [3]. Wuppertal Institute has published an analysis of common characteristics and site-specific differences of four coal-dependant European regions: Silesia PL, western Macedonia EL, Aragon ES, and Lusatia DE [9]. The analysis is focused on the distinction between the hard coal-producing region of Silesia and the lignite-producing regions. Silesia exhibits a high level of urbanisation. It is also strongly industrialised, with many sectors having considerable shares in regional GDP. On the contrary, lignite-producing regions remain rural areas with low population densities, where mines and thermal power plants are the main providers of jobs [9].

In this context, the Greek government announced the abolition of lignite-based electricity generation by 2028 and the increase in electricity generation via RES to 35% by 2030. Under the new strategy, there will be an ambitious program to accelerate the reduction of lignite-based power generation over the next decade [10]. Given the fact that 69% (2019) of Greece's lignite production takes place in the region of western Macedonia, this area will be at the heart of the energy transition, with all the threats but also opportunities that this entails [7,10].

Taking into account the fact that the climate mitigation is a collective effort in Europe, it seems fair that coal/lignite mining regions should receive support to master the challenges of this transition [2,3]. The Greek government is currently preparing a new comprehensive action plan for this purpose [10]. An essential part of this plan are the contributions granted via various European resources, such as the Fair Transition Fund and other financial mechanisms. However, the success of this plan relies on the decisions made by local and regional authorities regarding the development of new sustainable economic activities in the post-mining era [11]. Although actions aiming at the diversification from existing economic activities depending on lignite are necessary, first and foremost, the development of the optimum mix of land uses on the reclaimed mining areas is critical for achieving sustainability targets at the local and regional level.

The main objectives of this contribution are the determination of a series of sustainability criteria and indicators and, to a next stage, the development of an evaluation process for various land rehabilitation scenarios. This process is validated in the case of the western Macedonia lignite centre, a complex of surface mines that was the main pillar of the energy sector of Greece for more than six decades. More specifically, the paper is organized as follows: The second section provides background information on the evolution of the lignite industry in the western Macedonia region, the current situation and the perspectives of the local economy during and after the transition to a zero-lignite era, and the results of the mine land reclamation works carried out so far. The third section proposes a series of land uses that are considered to fit better in the examined case for amortising decarbonisation stresses, based on the circular economy principles and the founding tools available in the frame of the European Just Transition Mechanism. The fourth section describes a procedure for the evaluation of various land rehabilitation scenarios based on a PEST analysis and an optimisation algorithm that takes into account the criteria of revenues, investment cost, conservation of nature, and equity. In addition, the third section and the initial paragraphs of the fourth section provide references supplementary to these presented in the introduction, which strengthen the author's claims and suggestions. Finally, the last two sections provide a short discussion and conclusions.

2. Background

2.1. The Lignite Mines of Western Macedonia

Starting in the early 1950s, the lignite industry has critically shaped the development of western Macedonia. The decision to intensify the exploitation of domestic lignite deposits has been a central political option, supported by all Greek governments over the years. Up to now, 1.7 billion tons of lignite have been produced and more than 8.5 billion cubic metres of rocks have been excavated from four surface mines (Figure 1). The remaining exploitable reserves in the lignite field under exploitation by Public Power Corporation SA are estimated to be 820 million tons, while private mines reserves are 120 million tons. Public Power Corporation SA also has the rights to exploit 460 million tons of lignite located in areas where mining activities have not been developed so far [12].

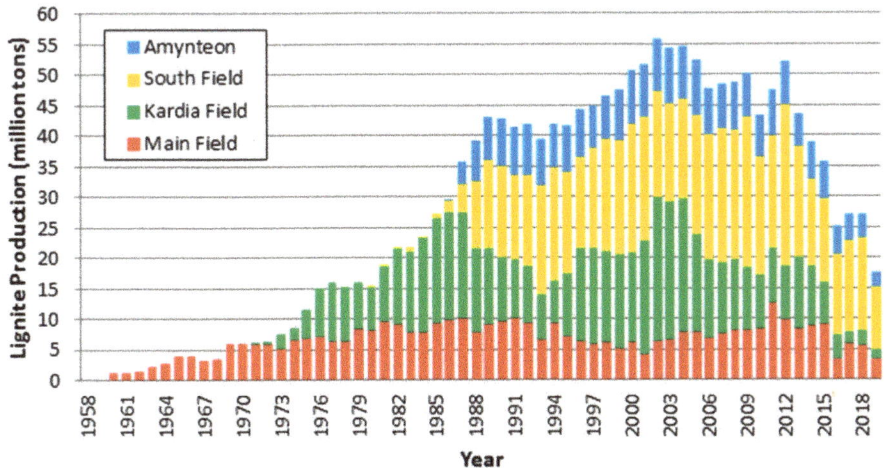

Figure 1. Lignite production of the western Macedonia Lignite Centre [12].

The lignite deposits of western Macedonia belong to an elongated sedimentary basin with a length of 250 km, which extends to NW into the SW territories of north Macedonia (Figure 2) [13]. The basin is divided into two elongated grabens with different stratigraphic evolution and subsurface morphology. The basin formation occurred at the end of the Tertiary era and its creation is considered to be a consequence of subsidence in large NW–SE fault zones [14,15]. The different sedimentation rates of the basin resulted in frequent intercalation between the lignite layers and the sterile, which consist, mainly, of marls and subsequently from clay and sand. As a consequence, the lignite deposits are characterised by a multiple-layered form, which deteriorated considerably the quality characteristics of excavated lignite compared to these of geological lignite due to the unavoidable co-excavation of sterile. The basement underlying the sedimentary basin includes Paleozoic schists, ophiolites, and granites. Above the basement lays the Pelagonian Structural Zone, which consists from Mesozoic dolomitic limestones, interlaying with volcanic sediments with ophiotic blocks, and flysch. The basin itself consists from Tertiary and Quaternary sediments with a maximum thickness of 1000 m. The upper part of the basin is filled with Miocene to Pliocene sand, sandy clay, lignite, and marl, mainly of fluvial to lacustrine origin. More specifically, in the surface lies the Quaternary sediments, then the Plio-Pleistocene unit that has a thickness of 20–100 m and consists of sand and clay, which intercalate with marls and conglomerates. Below this sequence lies the Plio-Miocene deposits consisting of layers of lignite and sand. These sediments are rich in $CaCO_3$ since it can be found in all the sediments of the sequence, including the lignite [14].

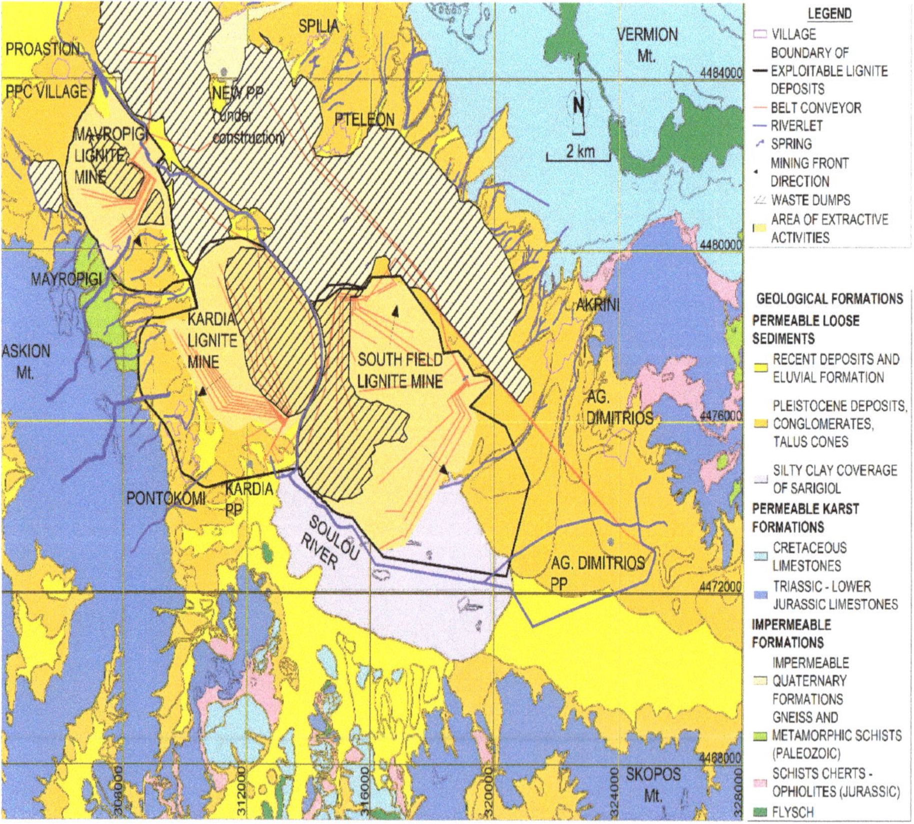

Figure 2. Geological map of the western Macedonia Lignite Centre (Ptolemaida mines) [13].

Nowadays, the mines occupy 17,000 ha (Figure 3). The area that was gradually expropriated for mine development purposes was mainly agricultural. Forests covered limited areas, usually close to watercourses. Moreover, several villages that were located in close proximity to the mines have been resettled [12].

Figure 3. Satellite view of the western Macedonia Lignite Centre, with the red curves indicating the two mining areas: Amynteon (NW) and Ptolemaida (SE).

The mines have been equipped with 42 bucket wheel excavators, 250 km of belt conveyors and 17 spreaders, while shovels and trucks are used for the excavation and transportation of the hard rocks that are present in the overburden strata. These formations usually require blasting. In the last year (2019), 17.7 million tons of lignite and 130 million cubic metres of rocks were excavated from the four active open-cast lignite pits. This production rate is significantly lower than the record figures of 55.8 million tons of lignite (2002) and 332 million cubic metres of rocks (2005). The mines meet the fuel demand of 11 thermal power units with a total installed capacity of 3737 MW, which should be gradually reduced due to the EU decarbonisation targets that are fully adopted by the Greek government [12].

Regarding water management, this is a critical issue for the operation of the surface mines and the implementation of the environmental protection and land reclamation programme. For this reason, a complex system of seven pumping stations, which also serve as sedimentation ponds, and numerous wells, which are responsible for the drop of the groundwater table around the pits, have been developed [14]. A small river that flows into Lake Vegoritis receives surface water and groundwater discharges pumped from the mines. The river has already been relocated a few times for the needs of the expansion of the mine's exploitation. Under normal operating and weather conditions, the river receives annually more than 40 million cubic metres [14]. The quality characteristics of the water allow its use for dust depression in-pit, irrigation, or even for water supply purposes. Deviations from water quality standards usually refer to the concentrations of suspended solids and electric conductivity values. Problems with acidic drainages have never been detected due to the presence of minerals rich in CaO in the entire lignite-bearing basin [15].

2.2. Western Macedonia Decarbonisation Impacts

The lignite industry has critically shaped the development of western Macedonia. For many decades, more than 25% of the regional GDP (gross domestic product) and more than 22,000 direct and indirect jobs have been unilaterally supported by local lignite activity needs [16].

Based on Hellenic Statistical Authority data, the use of an input-output model methodology for the period 2000–2016 showed that there is a direct correlation between lignite production and jobs. For every million tons of lignite produced, 185 jobs are maintained in the mining-energy sector and 725 jobs are created in the local labour market; this is a ratio of 1:3.9. The correlation of annual lignite production rates and employment in the western Macedonia region is illustrated in Figure 4. Although the R^2 value of the linear regression line of total employment is not high, the number of jobs in western Macedonia in 2028, when the lignite output will be zero, is estimated to exceed marginally 70,000 (i.e., 24% compared to year 2014), unless strong impact mitigation interventions are implemented [17].

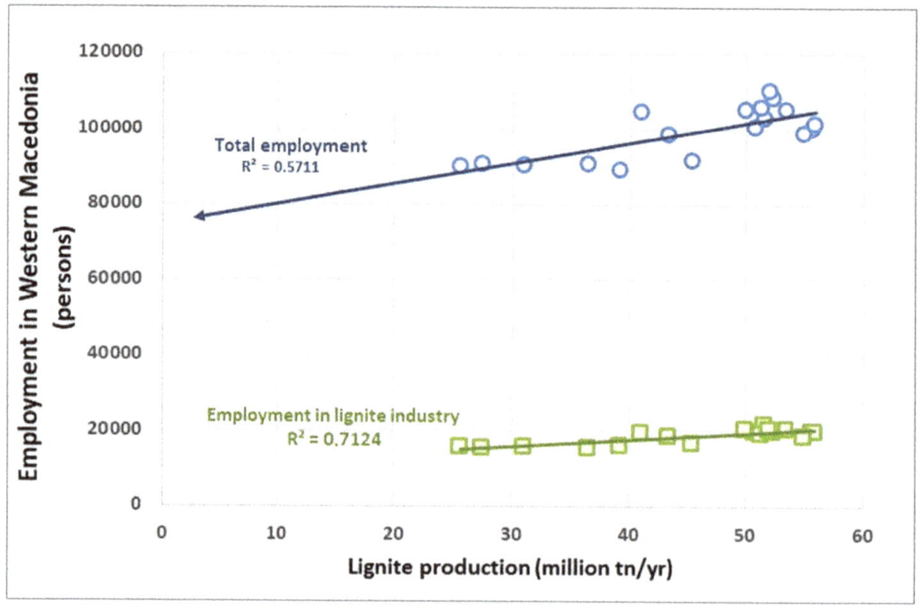

Figure 4. Lignite production and employment in western Macedonia.

Moreover, the regional GDP was very sensitive to the slight increase of lignite production during the period 2000–2005 (Figure 5). In the next years, when lignite production collapsed from 55.0 million tons in 2005 to 25.6 million tons in 2016, this correlation weakened (i.e., the slope of the blue regression line in Figure 5 is smaller than the one of the red line). This is probably due to deterioration of the stripping ratio, which kept the total rock excavations almost constant until 2013. The strong correlation of regional GDP with the excavated rock volumes is illustrated in Figure 6. In particular, this figure demonstrates how the regional economy is affected by the rock-moving works carried out by subcontractors of Public Power Corporation SA [16,17].

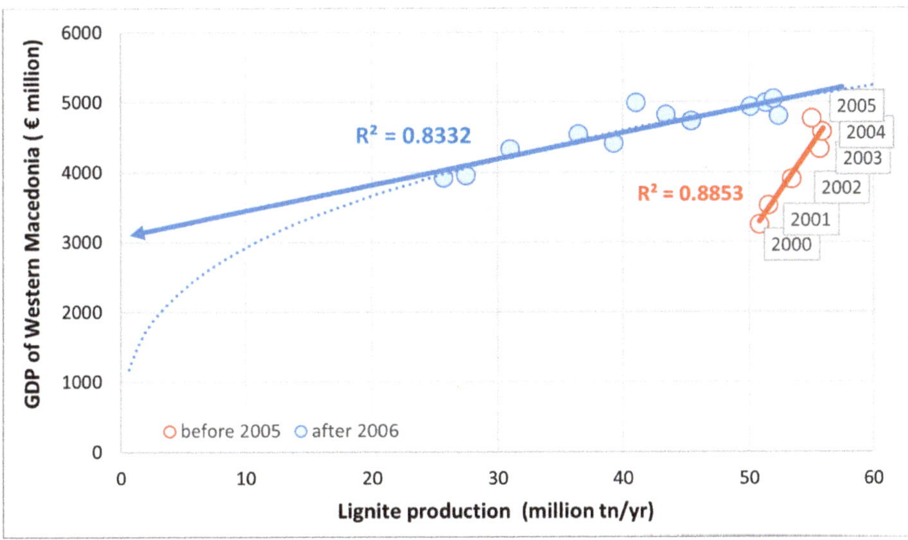

Figure 5. Correlation of GDP of the western Macedonia region and annual lignite production.

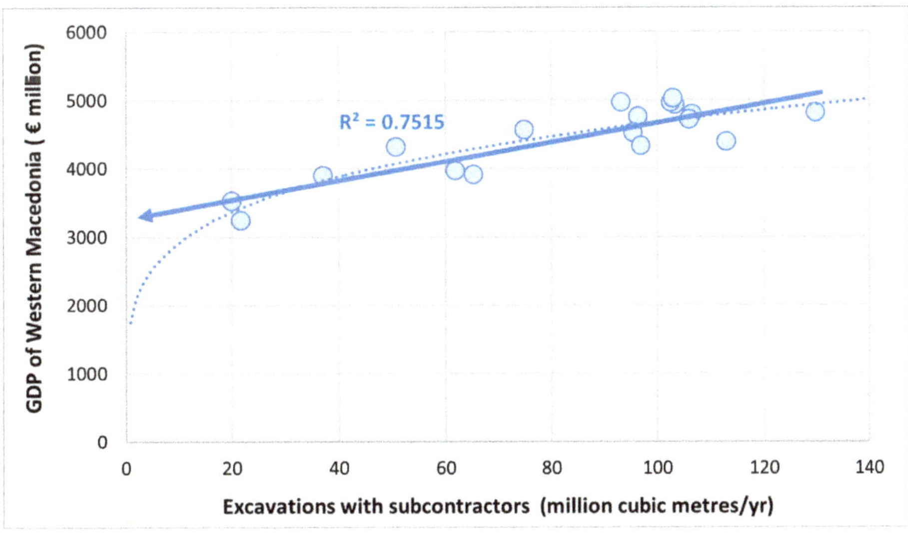

Figure 6. Correlation of GDP of the western Macedonia region and annual excavations carried out by subcontractors operating in lignite mines.

Based on the data presented in Figures 5 and 6, it can be predicted that, with zero lignite production and zero excavations in 2028, the GDP of west Macedonia will be moved between €3.2 billion (linear regression) and €1.5 billion (power regression). In the first case, if regional GDP is €3.2 billion, the wealth loss of the region of west Macedonia will reach 27%. In the second case, if regional GDP shrinks to €1.5 billion, urgent political intervention will be necessary to minimise demographic changes and to reduce economic and social impacts in the long term [17].

Summarising the effects of decarbonisation in the western Macedonia region and having as a reference the year 2013, the zero lignite production in 2028 is expected to bring the following consequences [11,16,17]:

- GDP decline by 26% and an annual loss of revenue of €1.2 billion.
- Loss of 21,000 jobs and an employment reduction in 24%.
- Total income loss for the period 2013 to 2028 of €9 billion.

Jobs loss and regional GDP reduction as a consequence of the withdrawal of lignite plants in the period 2009–2016 are presented in Figure 7, combined with a prediction for the year 2028, when lignite production is expected to be zero [17]. The above-presented Figures 4–7 make clear that western Macedonia's long-term dependence on the lignite cancelled every developmental effort, which was quantitatively reflected in low productivity diversification and low innovation rates. Moreover, it has created conditions that cannot be addressed by corrective interventions but require long-term productive restructuring policies, based on the competitive advantages of the wider region of west Macedonia. A key component of these policies should be the rehabilitation of the surface lignite mines.

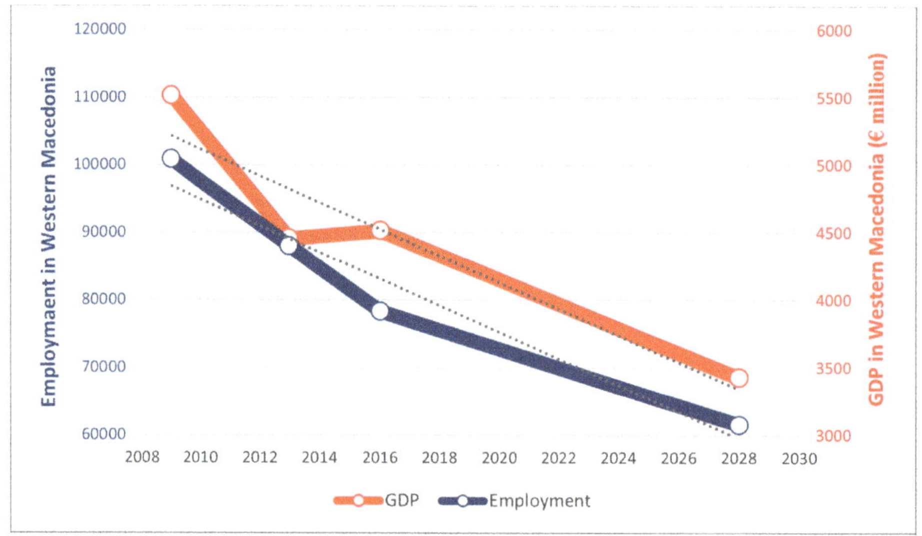

Figure 7. Forecast of GDP—employment progression in Western Macedonia.

2.3. Land Reclamation of Western Macedonia Lignite Mines

In Greece, all mining and quarrying activities operate according to the revised regulations determined in a ministerial decision amended in 2011. Moreover, mines and quarries have to meet quality standards and to apply preventive and mitigation measures that are described in numerous national laws and European directives [18].

Nevertheless, the main legal tool that regulates all environmental management decisions of a mining company is the environmental permit. The first permit referring to the mining activities at the western Macedonia lignite mining complex was signed in 2001, after a long period of negotiations with all the involved local and national authorities. Since then, additional permits have been signed for all mining operations, as well as for numerous auxiliary activities, such as the ash and asbestos cement disposal sites [18,19].

The main issue introduced by the permits is the implementation of a land reclamation programme according to specific guidelines dealing with waste heaps topography, landslides prevention, topsoil management, and reforestation [19]. Permits also refer to the costs of implementing the above-mentioned terms and conditions for the permitting period and until mine closure/rehabilitation (Table 1). This cost includes all the activities required for environmental management during mining operations and land reclamation according to the plans that have been approved by the authorities.

In addition, to compensate the adverse impacts of lignite mining on the environment and on other sectors of the local economy, the mine operator pays a revenue bond, which corresponds to 0.5% of its turnover. The total amount of money that has been paid from PPC so far (2018) is ca. €260 million and has been distributed to the prefectures, where lignite mining activities exist, accordingly to the produced lignite quantities. The regional authorities have taken full responsibility for selecting and implementing the development projects that are financed by this bond [18,19].

Table 1. Cost of implementing the terms and conditions of environmental permits [19].

Mine Complex.	Total Area Occupied (ha)	Environmental Cost until Permit Expiration (€)	Environmental Cost until Mine Closure—Reclamation (€)
Ptolemaida	14,792	25,000,000	90,000,000
Aminteo	5294	8,000,000	40,000,000

The development of a progressive land reclamation plan, which will be implemented throughout the entire life of the lignite mines, is a critical part of a successful land management strategy. Land reclamation ensures that the post-mining landscape is safe and is stable from physical, geochemical, and ecological perspectives; the water quality is protected; and a reliable environmental quality monitoring system has been installed [20]. Land reclamation is based on the map of land uses (Figure 8), which is contained in the Environmental Impact Assessment study that has been approved by the Ministry of Environment [20,21]. This map is subject to modifications when the permits are expired (usually every 10 years) and a new environmental impact assessment study is elaborated. The revised map of land uses may incorporate modifications relevant to the planning of the mines as well as changes of the type and spatial distribution of land uses.

Up to now, 3700 ha of mine land have been reclaimed. The usual reclamation practices include the development of farming lands in horizontal areas on the top of the waste heaps, and the reforestation of the sloped surfaces of the heaps' margins. The cultivated land is rented to local farmers and is seeded with wheat. The yields achieved vary considerably from site to site but, in general, are comparable to these reported for crops developed in the surrounding areas [22]. However, a study conducted by the Aristotle University of Thessaloniki showed that the development of agricultural activities in reclaimed mine lands cannot compensate for the job losses due to the mines' closure and power plants' decommissioning [23]. This situation does not change taking into account also the jobs related to honey and biomass production from reforested areas [24]. In this frame, for the near future, it is expected that PPC will continue the reforestation of sloped surfaces and the development of agricultural land in horizontal surfaces and, eventually, will start to construct photovoltaic parks in large horizontal plots of the waste heaps. Taking into account the topography of the pits and waste heaps and the hydrogeological conditions, it is estimated that 8% of the total area occupied by the mines will be covered by artificial lakes [21]. The remaining area can be used for livestock farming and reforestation, while 55% of this area is appropriate for agricultural activities and photovoltaic parks.

No matter what the development plans of PPC are and the commitments of the central government and regional authorities on the decarbonisation policies that will be applied in western Macedonia, the reclamation of the mine land is a legal obligation. In the following paragraphs, numerous arguments that prove the necessity of going with land reclamation one step further to land rehabilitation focused on the realisation of development programmes and innovative land uses that enhance the perspectives of the local economy are presented, among others.

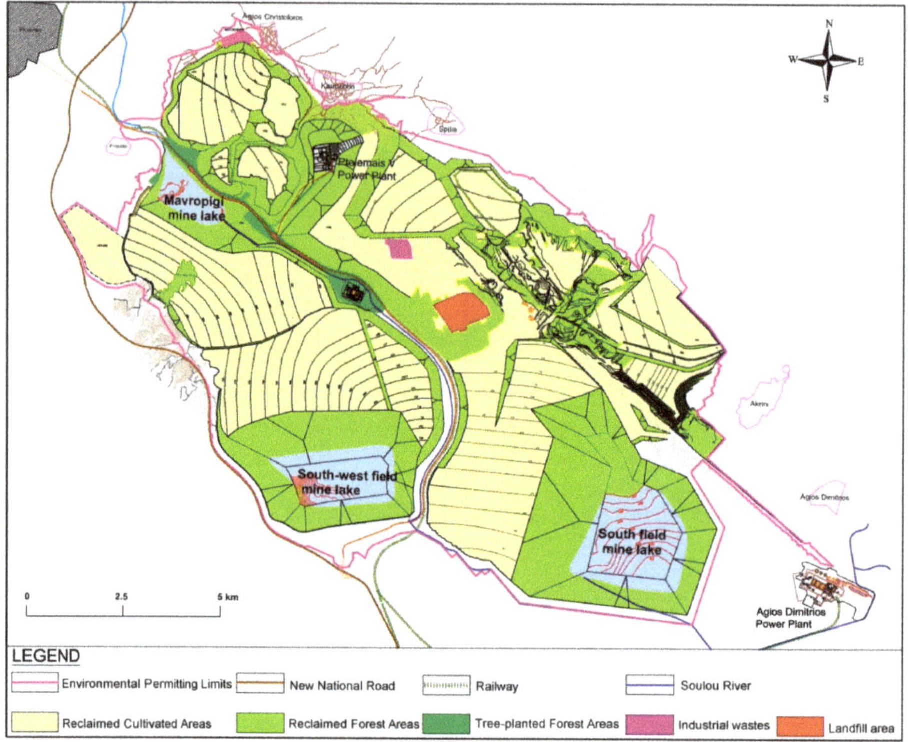

Figure 8. Land uses in the western Macedonia Lignite Centre after the mine closure and the completion of the land reclamation programme prescribed in the environmental permits [21].

3. Planning Rehabilitation for Amortising Decarbonisation Stresses

The mitigation of economic and social impacts of decarbonisation requires actions targeting beyond land reclamation. By adopting the terminology proposed by Lima et al. [25], from the 4R post-mining recovery practices (i.e., remediation, reclamation, restoration, rehabilitation), rehabilitation fits better in the case of western Macedonia lignite mines, where optimisation of land capacity for human use is a prerequisite for considering the socioeconomic prosperity of the region in the post-mining era, maintaining at the same time the ecosystem value. This is also in accordance with the society requirements, as these have been expressed by local and regional authorities and numerous stakeholders [26].

A fundamental target of rehabilitation is to create the best possible land capabilities that can enable a multitude of land uses in the future. In this frame, mine site rehabilitation should be planned to meet the following key objectives: (i) The long-term stability and sustainability of the landforms, soils and hydrological pattern of the site, (ii) the partial or full restore of ecosystem capacity to provide habitats for biota and services for people, and (iii) the prevention of pollution of the surrounding environment [26,27].

Particularly in the case of western Macedonia lignite mines, the vision of mine closure until 2028 urgently requires a rehabilitation strategy to guide all decisions and actions during the remaining mines life. These actions needs to take into account the following: (i) The public health and safety cannot be compromised; (ii) the natural resources are not subject to physical and chemical deterioration; (iii) the post-mining land uses may be sustainable in the long term; and (iv) the socioeconomic impacts need to be minimized; (v) opportunities, such as the maintenance of infrastructure, facilities, and services, need to be taken into account to maximise socioeconomic benefits [28].

Closure and rehabilitation planning must include the participation of community members, local and regional government, and development partners. In addition, mining companies must participate in local economic development initiatives to reduce the dependency of local communities on mines' operation before their closure [29,30].

It is worth noticing that unplanned closures create significant problems for the mining company, the community, and the regulator. In Australia, 70% of the mines have closed for reasons other than exhaustion of reserves, such as low commodity prices, high costs, safety or environmental breaches, policy changes, community pressures, closure of downstream markets, floods, and geotechnical failures [28]. This actually happens nowadays in western Macedonia, where the closure of lignite mines came 25 years earlier as a result of the high lignite extraction cost due to the EU regulations for CO_2 emission allowances.

3.1. The Circular Economy Concept

Considering the basic concept and priorities of the circular economy related to resource efficiency and material use, a circular economy analysis of post-mining land uses in surface mining operations could contribute significantly to investigating the long-term economic viability and sustainability of such projects. The enormous size and long duration of these operations allows the incorporation of several aspects of sustainability and circular economy flows to the land rehabilitation strategy, which has been applied for many decades in parallel to mines' exploitation as soon as some parts of the mine area are no longer needed for the development of pits and waste heaps (Figure 9). For instance, the conservation of topsoil cover, which is excavated separately in order to be spread to form a fertile reclaimed mine land, is a legal requirement that clearly affects the sustainable future of the greater mining area. Moreover, a series of measures for surface runoff and groundwater management, wastewater treatment, management, and disposal of solid and special waste shape the performance of every surface mine in terms of achieving the sustainability and circular economy goals [31].

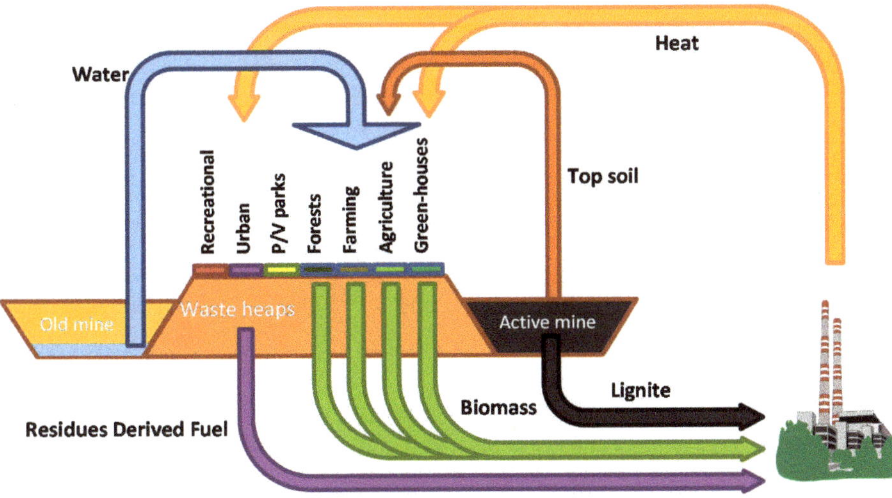

Figure 9. Schematic representation of circular economy practices applicable in surface lignite mines.

3.2. The Proposed Land Uses for the Rehabilitation of Western Macedonia Lignite Mines

Land resources are limited and finite. There is therefore a need to match land types and land uses in the most rational way possible so as to maximise sustainable production and satisfy the diverse needs of society [32]. Sustainable spatial development takes into consideration all environmental, social, economic, cultural, institutional, regulatory, and demographic aspects. Planning involves anticipation of the need for change as well as reactions to it [29].

If surface mine rehabilitation is planned, the scenarios of land uses must aim at outcomes that are climate smart, adaptable, and resilient. Successful rehabilitation is the state in which land is sustainable in the long term and can support agreed post-mining land uses, it can be effectively managed after the mine closure, and residual and latent risks are minimised. These risks usually imply the inability to maintain defined land capabilities due to loss of available soil, settlement, loss of surface water yield and loss of vegetation, lack of long-term water availability, and lack of maintenance of retained infrastructures by the new owners [26].

Kivinen [33] examined post-mining land uses in 51 metal mining sites in Finland. Forests cover 75% of the total post-mining area. An additional 17% is covered by water bodies (lakes, rivers, and wetlands). The agricultural land accounts for 8%, while sites relevant to commercial, recreational, or national heritage activities have been developed only in mines located in densely populated territories.

In the western Macedonia region, where lignite mining and electricity generation operations provide the only significant mainstream economic activity, post-mining repurposing scenarios and decisions about land use planning must refer to the greater mining area or even to the entire region. For instance, the development of certain crops in reclaimed waste heaps can be proved as a waste of money, when the same crops can be productive in neighbouring sites that were not affected by mining operations.

In this frame, the following land uses have been proposed by various stakeholders for the post-lignite era:

Agriculture (extensive farming): The crops of cereals are the typical agricultural activity in the greater lignite-bearing basin. Although farmers see mining land as inferior, they demand from the mine operator to rent large plots in order to extract benefits from them unsustainably, taking advantage of the financial support provided by the EU for certain cultivations. The development of modern farms based on the systematic agriculture principles is the ideal future, provided that water reservoirs and irrigation systems will be constructed.

Agriculture (green houses): It is the most promising type of agricultural activity, provided that cheap thermal energy produced by nearby located power plants should be supplied to farms.

Livestock farming: It is a traditional economic activity in a few of the communities located at the margins of the basin. It does not attract the interest of young farmers although Greece's import–export balance for meat and dairy products is negative.

Forests: Taking into consideration that the woodland areas destroyed due to the mines' development were very limited and in accordance to the rehabilitation principles, which aim at the reinstatement of ecosystem functionality and land productivity, allowing a different species composition from the original ecosystem, the plantation of various coniferous and deciduous trees is evaluated based on criteria related to aesthetics, slope stability, and revenue (e.g., biomass production for fuelling nearby located thermal plants, honey production). The new ecosystem may be simpler in structure than the original but more productive [27].

Recreational activities: They can be developed in reforested areas and around water bodies (final mine pits) located in the proximity of towns and villages, provided that appropriate infrastructures of a relatively low cost should be constructed. However, it is difficult to attract visitors from other regions or from abroad due to the competition from sites that have a reputation of providing high-quality tourist services. Nevertheless, an issue that probably requires special attention is the development of activities relevant to the industrial heritage.

Photovoltaic parks: Their development is considered by the lignite mine's operator as a feasible option for switching from a carbon-intensive electricity generation portfolio to a 'green' future. The availability of land and infrastructures relevant to the connection with the electricity transport network far outweigh the increased operating costs due to the longer distance from consumers.

Urban: Taking into account the willingness of local people to relocate to the major cities of the region, the development of residential areas seems to have limited interest. On the contrary, the location of industries and/or public sector activities, such as universities, research institutes, and hospitals, is considered historically as a measure for mitigating demographic decline and supporting the development of new sectors of strategic importance. In addition, the 'success story' of the construction of a landfill and waste management plant with a capacity of 120,000 t/year, which processes all the solid waste of the western Macedonia region, should act as a pilot project for the development of other waste-processing activities.

Taking into consideration that, after the phase out of the last thermal power plant in 2028 the remaining exploitable lignite reserves are estimated to be 500 Mt, surface mining can be another land use. The produced lignite can be used to non-energy related markets (e.g., activated carbon) or can be co-combusted in thermal power plants that use biomass or residues-derived fuel as basic fuels. It is worth noticing that in Finland, the re-opening of metal mines that have been closed in the period 1924–2016 was planned or under development in five cases (ca. 10%) and exploration permits were applied or admitted for half of the post-mining areas [33].

3.3. A Just Transition Mechanism

The required restructuring policies for revising the plan of land uses and restructuring the economy of the western Macedonia region can be part of the Sustainable Europe Investment Plan, which will mobilize more than €1 trillion of private and public investments over the next decade. This plan will support climate-neutral investments related to environmental protection and social cohesion [34].

This transition to a new economic development model will require large investment and a decisive policy response at all levels. Although all regions will need funding, the transition will imply a significant challenge to some of them, which have to adopt radical restructuring of their economies and to develop new economic activities and workforce skills. Regions highly dependent on coal/lignite utilisation for electricity generation purposes will need to ensure that business innovation and activities diversified from the mining and thermal power generation can maintain the social cohesion and provide an adequate number of new jobs.

To address the site-specific challenges encountered by the most impacted regions, the European Commission proposes a Just Transition Mechanism that provides dedicated support to generate the necessary investments. This mechanism will consist of three pillars, which will offer different grant and financing tools in order to cover the various types of support required by the public and private sectors of the impacted regions. These pillars are:

- The Just Transition Fund, which will provide primarily grants;
- The InvestEU, which will manage private investments; and
- A new public sector loan facility for supplementary investments to be leveraged by the European Investment Bank.

All programmes that will be applied in the frame of the Just Transition Mechanism will be accompanied by dedicated advisory and technical assistance [34].

4. Evaluation of Mine Rehabilitation Scenarios

Land management must explicitly consider what are the optimum land uses for a particular mine site and, at the same time, what are the optimal sites in an area or region for developing particular land uses. The proposed suitability criteria include site environmental characteristics, infrastructures, environmental hazards, development impacts, and institutional concerns [35].

Rehabilitation implies a decision-making process, which assesses the impacts of various land uses on the living environment and ecosystem and, to a next stage, the potential of these land uses to boost economic development. Land use planning, in turn, is a systematic evaluation of the land and water potential, in both qualitative and quantitative terms, and the economic and social conditions in order to select and implement the best land use options that meet the needs of people while safeguarding natural resources for the future and diminishing environmental risks [32,36].

As far as the conservation of nature is of concern, the Society for Ecological Restoration recommends the use of nine ecosystem attributes for measuring mine rehabilitation success: (i) Similar ecosystem diversity and community structure to those of reference sites, (ii) presence of indigenous species, (iii) presence of a functional group of species that are necessary for long-term stability, (iv) capacity of the environment to sustain reproducing populations, (v) normal functioning, (vi) integration within the topography, (vii) elimination of probable threats, (viii) quick recovery in cases of natural disturbances, and (ix) self-sustainability [37].

Furthermore, mining industry interferes in many ways in the social affairs. This implies a commitment to minimise the adverse impacts of mining on communities located in close proximity to the mines, and also introduces the issue of how to maintain the prosperity and sustainability of the society, which is closely related to the developing of processes and structures that support the potential of current and future generations to create healthy and liveable communities [38].

Rehabilitation planning usually includes comprehensive spatial analysis of the mining area, including the already reclaimed mine land. Thus, the literature abounds with papers proposing methods and techniques that support decision-making for the spatial distribution of alternative land uses based on a series of criteria, most of them by incorporating GIS-based applications.

For the lignite mines of western Macedonia, Pavloudakis et al. developed a methodology in order to match the land characteristics and repurposing scenarios of mine land rehabilitation based on the following broader categories of criteria [39]: (i) Location: Proximity to lakes, archaeological sites, and residential areas; (ii) geotechnical stability; (iii) topography; (iv) soil fertility: pH, acidity, alkalinity, concentration of nutrients, mechanical properties, and access to irrigation systems; and (v) environmental risks: Metal concentrations in soil and water. Palogos et al. [40] proposed an alternative approach by adding evolutionary algorithms to the land use evaluation process. Furthermore, Wang et al. [41] proposed a reclamation strategy for a mining site in Liaoning Province, China, combining land suitability analysis and ecosystem service evaluation methods. Forest, agriculture, and developed land were examined as alternative land uses using land suitability criteria related to topography, climate and socioeconomic factors, and targeted ecosystem services material production (e.g., timber), water conservation, soil protection, carbon sequestration, oxygen release, and air purification.

In this context, in order to emphasise the social and economic issues that must be taken into consideration in decision-making processes relevant to mine land rehabilitation, in Table 2, a comparative PEST (i.e., political, economic, social, technological) analysis is presented considering seven alternative land uses that are possible to be developed after the closure of the surface lignite mines of western Macedonia.

From the land uses analysed in the previous paragraph and in Table 2, green houses, and recreational and urban/industrial areas will occupy a relatively small percentage of the 17,000 ha of reclaimed mine land that will be available after the mines' closure. Thus, the development of these land uses does not jeopardise other uses and, to the extent that they help the achievement of the objectives of the land rehabilitation strategy, they must be politically and financially supported.

Furthermore, in order to optimise the total area to be covered by each of the four remaining land uses: Extensive agriculture, livestock farming, forests, and photovoltaic parks, an optimisation algorithm was developed. The algorithm differs from the decision-making methodology presented by the authors of this paper in previous studies [39,40] since it is focused on the selection of the optimum mix of land uses that meets certain environmental, economic, and social targets relevant to the rehabilitation of the mines without paying attention to the spatial distribution of land uses. In other words, this approach is driven by the goal to be achieved, which is the long-term prosperity of the society, and not by the restrictions associated with a specific set of criteria, assuming that financial resources are available for complete or even partial lifting of these restrictions.

Table 2. PEST analysis of the major macro-environment factors that affect decisions regarding post-mining land uses [11,23,27–29].

Land Use	P Political	E Economic	S Social	T Technological
Agriculture	Some crops are regulated by state or EU directives; establishment of cooperatives, etc.	Low productivity land; low investment cost; expensive irrigation	Traditional economic activities in the greater mining area; moderate number of new jobs	Topsoil management and soil remediation are required
Green houses	Lack of regulatory framework for heat supply to agro-sector; investment support mechanisms are required	Moderate investment cost highly competitive market; considerable reduction of minimum viable acreage compared to extensive farming	The large number of agronomists working in the area makes easier the know-how transfer	Some techniques (e.g., hydroponics) do not require soil remediation; need to develop a marketing strategy to grab a big market share
Livestock Farming	Establishment of cooperatives	Ideal for low productivity lands; moderate to high investment cost for vertical growth of production (from forage to dairy products)	Limited number of new jobs; traditional activity for some villages in the greater area	Soil remediation is required
Forests	Limitations arising from the forest law and State planning	Possibility of forest exploitation for timber, biomass and honey production; careful calculation of the logistics costs is required	Moderate number of new jobs; upgrading of living conditions	Limited number of tree species can grow under the certain weather and soil conditions
Recreational activities	Limitations concerning safety of visitors (e.g., access to lakes)	Local population is not enough to guarantee economic viability; investments required for development of attractions	Most of them create limited number of new jobs; upgrading of living conditions; connected with industrial heritage	Regular monitoring of soil and water quality and slope stability is required
P/V Parks	Pending approval of permits	Availability of low-cost land and infrastructures for connection to electricity network	Have already been criticised for the limited number of new jobs during operation; limited argues for degradation of landscape, etc.	Opportunities for know-how development and creation of new businesses related to the construction of P/V parks, panels, etc.
Urban (industrial)	Establishment of industrial area; investments support decisions are required; construction works must follow technical specs.	Large investment cost; insufficient transportation networks (railways, airports)	Large number of new jobs; training—education of personnel is required	Need for know-how transfer; opportunity for development of new supporting activities

The proposed optimisation algorithm is based on the following broader categories of criteria [32]:

Revenues: Land use must be economically viable. The basic objective of land use planning is the efficient and productive use of the land by matching different land uses with the areas that will yield the greatest benefits at the minimum cost.

Investment: Although significant preparatory actions are underway to develop financial tools to support the transition of the regional economy, the minimisation of the required investment is still assessed positively.

Conservation of nature: In order to ensure continued production in the future, the conservation of the natural resources on which that production depends on is beyond any negotiation.

Equity: Land uses must be diverse and socially acceptable. People with different education and skills must have equal opportunities of employment. Poverty and social exclusion must be eliminated.

For each one of the four categories of criteria (or just criteria, for simplification reasons) j and for each land use i, a rate R_{ij} is given (using a 1–10 scale) Each of these rates is calculated as the average of the rates given by 10 experts (mining engineers, agronomists, and surveyors), who are or were involved in land reclamation projects carried out by PPC in the surface mines of western Macedonia region (Table 3).

Table 3. Rating of land uses for each of the four examined criteria.

Criteria (j):	Land Uses (i):			
	1. Agriculture	2. Farming	3. Forests	4. P/V Parks
1.Revenues (max)	6	6	3	10
2. Investment (min)	6	10	8	2
3. Conservation of nature (max)	5	7	10	5
4. Equity (max)	8	4	4	6

Then, the team of experts and the stakeholders involved in the land use planning process decide about the threshold of the rates that must have the combination of land uses regarding each one of the four criteria. In the examined case of the lignite mines of the western Macedonia region, the threshold values that were determined by the group of PPC experts are given in Table 4.

Table 4. Threshold values for each one of the examined criteria of land use planning.

Criteria (j):	Threshold Values TVj
1.Revenues (max)	7
2. Investment (min)	5
3. Conservation of nature (max)	7
4. Equity (max)	6

Then, the scores (average rates) of 286 combinations $[\sum_{n=1}^{11} n(12-n)]$ of land uses were calculated assuming that these four uses will cover 100% of the available land. For each land use *i*, the land coverage *Pi* ranged from 0% to 100% with a step of 10%. The optimum combination of land uses (i.e., the optimum percentage of area coverage for each one) is determined based on the sum of deviations (D) of the average rates from the determined threshold values for all the four criteria. The modelling equation has the following form (Equation (1)):

$$D = \sum_{j=1}^{4}\left(if\ \sum_{i=1}^{4}(Rij.Pi) - TVj < 0\ then\ \sum_{i=1}^{4}(Rij.Pi) - TVj\ else\ 0 \right), \tag{1}$$

where:

D, the deviation of a certain combination of land uses coverages from the optimum score;

i, the land uses;

j, the land uses assessment criteria;

R_{ij}, the rate given to the land use i based on the criterion j;

P_i, the land coverage of the land use i, expressed as percentage % of the total mine area available for rehabilitation planning

TV_j, the threshold value for the criterion j.

For instance, if the land coverages of agriculture, farming, forests, and P/V parks have the percentages 10%, 20%, 40%, and 30%, respectively, and using the threshold values of Table 3, the above formula has the following form:

$$
\begin{aligned}
D = {} & 6 \times \frac{+6 \times}{} \quad 10\% \quad \frac{+3 \times}{40\%} \quad \frac{+10 \times}{30\%} \quad -7 \quad = \frac{6.0}{-7} \quad = \frac{-1.0 <}{0} \rightarrow -1 \\
& 6 \times \frac{+10 \times}{} \quad 10\% \quad 20\% \quad \frac{+8 \times}{40\%} \quad \frac{+2 \times}{30\%} \quad -5 \quad = \frac{6.4}{-5} \quad = \frac{+1.4 >}{0} \rightarrow 0 \\
& 5 \times \frac{+7 \times}{} \quad 10\% \quad 20\% \quad \frac{+10 \times}{40\%} \quad \frac{+5 \times}{30\%} \quad -7 \quad = \frac{7.4}{-7} \quad = \frac{+0.4 >}{} \rightarrow 0 \\
& 8 \times \frac{+4 \times}{} \quad 10\% \quad 20\% \quad \frac{+4 \times}{40\%} \quad \frac{+6 \times}{30\%} \quad -6 \quad = \frac{5.0}{-6} \quad = \frac{-1.0 <}{0} \rightarrow -1
\end{aligned}
$$

$$D = -2$$

The above-described approach identifies the combinations of land uses that achieve a more balanced development, trying to meet at the same time the standards that have been set for economic prosperity, rational use of funds, nature conservation, and equity. Therefore, it ignores how much above the threshold values are the rates of the criteria for a certain combination of land uses and takes only into account the rates that are below the threshold. For instance, the D value decreases considerably, taking the value −3, if the rate for a certain combination of criterion and land use is 4 and the threshold is 7, but it does not increase at all (i.e., it has a zero value) if the rate for another criterion and for the same land use is 9 and the threshold is 6.

The optimum percentage ranges for the examined land uses in the case of the threshold values presented in Table 4 is shown in Figure 10a. The combination of land uses with the best score (rate) is indicated by the red dots and the percentage ranges correspond to the upper 5% of the scores. According to these results, it seems that the opinions of PPC experts regarding the selection criteria of land uses favour P/V parks (40%) instead of agriculture (30%) as a high revenue activity, although it requires a higher cost of investment. Forests (20%) and livestock farming (10%) hold smaller shares that are necessary for the preservation of natural features and the biotic environment. Taking into account the fact that 55% of the available area is appropriate for agricultural activities and P/V parks, as it has already been mentioned in paragraph 2, the percentage of 70% for these two land uses can be achieved only if the plantations targeted for biomass or biofuel production will be extended to sloped surfaces located in the perimeter of waste heaps.

In Figure 10b–d, the results achieved by using three different sets of threshold values are illustrated. By increasing the threshold value for revenues from 7 to 9 and keeping all the other thresholds constant, the percentages of P/V park areas and farming land increase and those of forest and agricultural land decrease. If the conservation of nature is the predominant target of the land reclamation strategy, receiving a threshold value of 10, the land coverage of P/V parks increases further and the forest land percentage increases at the expense of farming land. Finally, if the stakeholders favour not only the conservation of nature but also the cost reduction, even if the potential for higher revenues is deteriorated, the forest coverage increases dramatically, and all the other land uses are reduced.

The optimum land uses based on the rates of Table 3 and assuming equal weighing factors for all the criteria (j) are shown in Figure 11. This case implies that threshold values are neglected or, alternatively, the proposed algorithm is applied using the threshold value of 10 for all the criteria. As it was expected, the maximum score corresponds to 100% of land coverage by the land use that exhibits the higher average score for the four criteria (i.e., farming with average score 6.75/10). This approach does not favour diversification of land uses and this did not change if different weighing factors were used for the examined criteria. Moreover, in order to check further the sensitivity of the proposed criterion for land use coverage, in Figure 11, the results are shown, which correspond to a set of threshold values equal to the average score of each one of the examined land uses. This approach must also be avoided because it obviously has the tendency to distribute equally the land coverage percentages. Finally, in the last two sets of columns of Figure 11, the land coverage results are presented, using the threshold values of 6 and 7, respectively, keeping them constant for all the four examined criteria. The remarkable difference resulting from this comparison indicates that the higher the threshold values used, the higher the influence of the high scores achieved for certain combinations of land uses criteria, and the lower the threshold values, the higher the influence of the low scores (in fact, the scores above the threshold value are neglected).

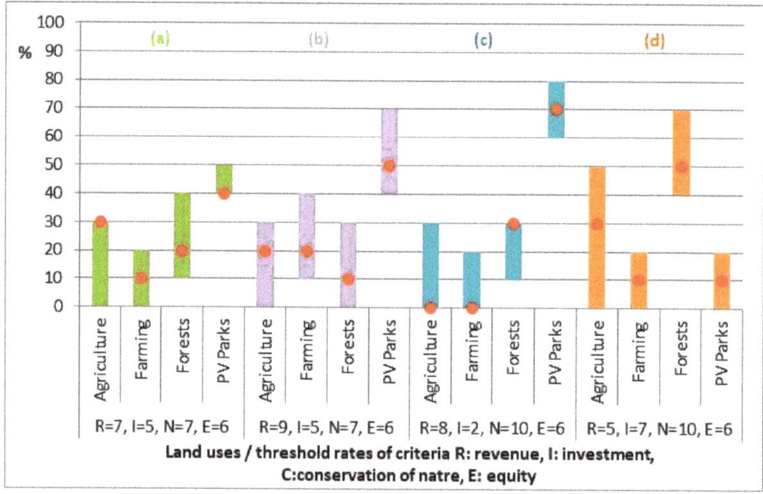

Figure 10. Ranges of land uses' percentages for different sets of threshold values for the examined optimisation criteria.

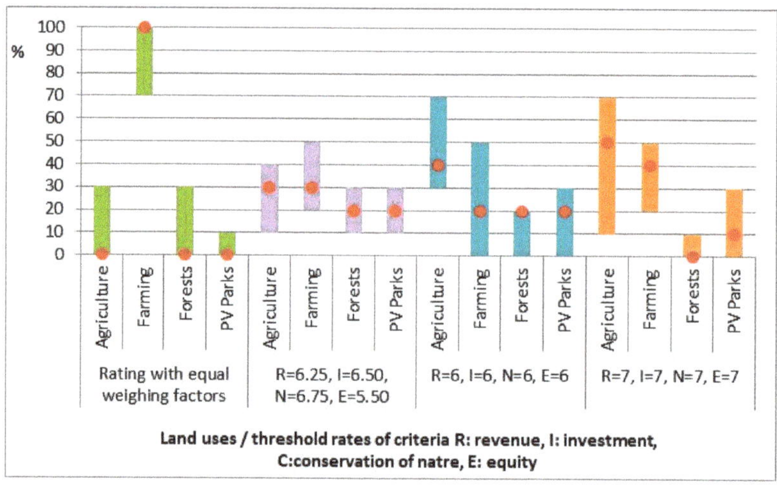

Figure 11. Sensitivity analysis of land uses' percentages in relation to various sets of threshold values for the examined optimisation criteria.

5. Discussion

In order to amortise decarbonisation stresses, the western Macedonia region needs, among others, an innovative mine land rehabilitation strategy. Taking into consideration the basic concept and the priorities of the circular economy, the present paper proposes seven land uses: Agriculture (extensive), green houses, livestock farming, forests, recreational areas, photovoltaic parks, and urban (industrial). All of them support the circular economy concept and each one complements the others in order to achieve sustainability targets. Urban development, which includes in a broader sense industrial and institutional development, is called for mitigating the economic and social impacts caused due to the loss of employment. Other uses, such as agriculture, livestock farming, and forests, can contribute to the energy economy cycle in numerous ways: Supplying an alternative fuel, which can co-combusted with lignite in thermal power plants; utilizing waste heat produced from thermal power plants; and utilizing water stored in the artificial lakes that will be developed in the final mine pits.

- The PEST analysis and the basic scenario that was examined with the decision-making algorithm clarified the expectations of the society from the different land uses.
- Extensive agriculture is an option that concerns only a few professional farmers while green houses are a promising investment, provided that the current regulatory framework will change and heat wasted by the thermal power plants will be available for uses other than district heating.
- Further to the conservation of the nature, reforestation can contribute to the local economy only if it is targeted to biomass production.
- Recreational activities that have been proposed so far by numerous stakeholders vary from mountain biking and fishing to industrial heritage parks and car racing circuits. Of course, the outputs expected from each of those activities differ considerably and must be evaluated accordingly.
- Photovoltaic parks are a realistic alternative for the creation of a new high-revenue activity, taking into account that PPC has decided to diversify its energy sources portfolio and is willing to pay the cost of investment and to take the relevant risks.
- Urban (industrial) development is the only land use that can create a large number of new jobs.

As far as the proposed optimisation algorithms is of concern, it was developed having in mind that the long-lasting prosperity of the society is closely related to the development of a strategy that balances effectively between conservation of nature and economic growth. The rates and the threshold values that were given by the experts contacted by the authors reflect a tendency, which has been gradually developed in the last years in the face of the shrinkage of lignite exploitation activities, to treat equally both environmental protection and economic development objectives.

Furthermore, the threshold values that were set by the experts for the three of the criteria (7 for revenues, 7 for conservation of nature, and 6 for equity) were higher than the average rates given from them to the four land uses for each of those criteria (6.25 for revenues, 6.75 for conservation of nature, and 5.50 for equity). Only for the cost of investment the threshold value (5) was less than the average rate (6.50). This fact is probably related to the large expectations of the local authorities and people involved in land rehabilitation projects regarding a generous financial support through the Just Transition Mechanism.

6. Conclusions

From the perspective of the new growth strategy of the EU and the ongoing changes in the world energy market, the reduction or phasing-out of lignite mining and power generation activities is considered as one of the most cost-effective methods to achieve CO_2 emissions reduction.

In this context, the Greek government announced the abolition of lignite-based electricity generation by 2028. For the region of western Macedonia, where in the recent past the installed capacity of lignite-fired power plants exceeded 4300 MW, the transition to a sustainable low-carbon economy entails threats but opportunities. The required restructuring policies can be part of the Sustainable Europe Investment Plan, which will mobilize at least €1 trillion of private and public investments over the upcoming decade. The supported actions must aim at the diversification from existing lignite-depended economic activities as well as at the revision of the spatial plan of the entire region so as to exploit the productive potential of all economic sectors.

Focusing on the rehabilitation of the surface lignite mines, the selected land uses should create new jobs and revenue without compromising on the objectives of protecting the living environment and ecosystem. At the same time, land uses should be self-sustaining, which means they will be able to generate profits when funding from European or national sources will cease. Considering the basic concept and priorities of the circular economy related to resource efficiency and material use, it could offer a reliable methodology for assessing the long-term economic viability and sustainability of numerous projects.

The application of a PEST analysis in the examined case study of the surface lignite mines of western Macedonia show that a series of economic activities that are expected to create many jobs combine the advantage of small land requirements.

Examples of this type of land use, which should be supported both politically and financially, are:

-   The construction of industrial complexes. For instance, it has already been proposed the construction of factories that produce photovoltaic panels and batteries for electric cars.
-   The construction of industrial heritage parks, utilizing part of the building infrastructures of the mines and power plants.
-   The development of greenhouses that will use cheap heat supplied by the nearby located power stations.

For the land uses occupying large areas, i.e., agriculture, forestry, livestock farming, and photovoltaic parks, an optimisation algorithm is proposed for determining the mix of land uses that maximise the revenue, equity, and natural conservation and minimise the cost of investment. According to the results obtained based on the set of rates and threshold values determined by a group of PPC's land reclamation experts, photovoltaic parks seem to be a more attractive option than extensive agriculture, as regards land uses that are worth investing in, while livestock farming and forests are considered necessary to safeguard the ecosystem functions.

These results would be probably different if representatives of public and local authorities and other stakeholders were taking part in the group of experts that decided about the rates and the threshold values that were fed to the proposed algorithm. Nevertheless, the proposed algorithm can be used to support decision-making procedures that will take place in the near future in the western Macedonia region concerning the development of land rehabilitation plants and the rational distribution of funds coming from the Just Transition Mechanism.

The algorithm is also applicable to various decision-making problems, not necessarily related to surface mine rehabilitation, in cases that a group of experts and/or stakeholders have to determine the optimum mix of land uses, development project proposals, etc., each of which claims from the rest a share of a limited resource, in terms of space, funding, etc. To this extent, the algorithm can be used by every surface mine operator in order to develop an effective land rehabilitation plan.

Author Contributions: Conceptualization, F.P., C.R. E.K. and N.K.; methodology, F.P., C.R. E.K. and N.K.; validation, F.P., C.R. and E.K.; formal analysis, F.P., C.R., E.K. and N.K.; data curation, F.P., C.R. and E.K.; writing—original draft preparation, F.P., C.R. and E.K.; writing—review and editing, F.P., C.R., E.K. and N.K.; visualization, F.P., C.R. and E.K.; supervision, F.P., C.R. and E.K.; project administration, F.P. and C.R. All authors have read and agreed to the published version of the manuscript.

Funding: This research received no external funding.

Conflicts of Interest: The authors declare no conflict of interest.

References

1. European Commission. *The European Green Deal, COM (2019) 640 Final*; European Commission: Brussels, Belgium, December 2019.
2. Alves Dias, P.; Kanellopoulos, K.; Medarac, H.; Kapetaki, Z.; Miranda-Barbosa, E.; Shortall, R.; Czako, V.; Telsnig, T.; Vazquez-Hernandez, C.; Lacal Arántegui, R.; et al. *EU Coal Regions: Opportunities and Challenges Ahead*; EUR 29292 EN; Publications Office of the European Union: Luxembourg, 2018; ISBN 978-92-79-89884-6. [CrossRef]
3. Sartor, O. *Implementing Coal Transitions: Insights from Case Studies of Major Coal-Consuming Economies*; IDDRI and Climate Strategies: Paris, France, 2018; 42p.
4. Artur, R.-M. *A European Strategic Long-Term Vision for a Prosperous, Modern, Competitive and Climate Neutral Economy, COM (2018) 773 Final*; European Commission: Brussels, Belgium, November 2018.
5. European Commission. *Quarterly Report on European Electricity Markets, Market Observatory for Energy, DG Energy*; European Commission: Brussels, Belgium, 2019; Volume 13.
6. Carbon Tracker Initiative. *Apocalypse Now*; Carbon Tracker Initiative: London, UK, 2019.
7. DeCarb. *Ex ante Economic and Social Impact Assessment of Regions' Decarbonisation*; Methodology Report; DeCarb Interreg Europe—EU ERDF, Stara Zagora Regional Economic Development Agency: Stara Zagora, Bulgaria, 2018; 51p.
8. Campbell, S.; Coenen, L. *Transitioning Beyond Coal: Lessons from the Structural Renewal of Europe's Old Industrial Regions*; CCEP Working Papers 1709; Centre for Climate and Policy, Crawford School of Public Policy, The Australian National University: Canberra, Australia, November 2017.
9. Wehnert, T.; Bierwirth, A.; Buschka, M.; Hermwille, L.; Mersmann, L. *Phasing-out Coal, Reinventing European Regions—An Analysis of EU Structural Funding in four European Coal Regions*; Wuppertal Institute for Climate, Environment and Energy: Wuppertal, Germany, 2017.
10. Greek Ministry of Environment, Energy and Climate Change. *National Energy and Climate Plan*; Greek Ministry of Environment, Energy and Climate Change: Athens, Greece, September 2019.
11. Giannakopoulos, D.; Karlopoulos, E.; Mavromatidis, D.; Sidiropoulos, A.; Topaloglou, E.; Tourlidakis, A. Post-coal Economy and Structural Transformation—Project Lab.: Western Macedonia Case. In Proceedings of the Coal Regions in Transition Platform, Working Group Meetings and High-Level Dialogue on Financing and Investments, Brussels, Belgium, 26 February 2018.
12. Roumpos, C.; Pavloudakis, F.; Liakoura, A.; Nalmpanti, D.; Arampatzis, K. Utilisation of lignite resources within the context of a changing electricity generation mix. In Proceedings of the 10th Jubilee Intl. Brown Coal Mining Congress: "Brown Coal Today and in the Future", Bełchatów, Poland, 16–18 April 2018.
13. Louloudis, G.; Mertiri, E.; Liakoura, E.; Kasfikis, G. *Hydraulic Protection Works of Ptolemaida Mines from Surface Waters*; Public Power Coroporation SA, Sector of Hydrogeological Studies: Athens, Greece, July 2020. (In Greek)
14. Koukouzas, N.; Karametou, R.; Gemeni, V.; Roumpos, C.; Sachanidis, C.; Pavloudakis, F. Improvement methods of pumping stations function in coal mines of Western Macedonia Lignite Centre. In Proceedings of the 14th Intl. Symposium of Continuous Surface Mining, Thessaloniki, Greece, 24–26 September 2018.
15. Sachanidis, C.; Pavloudakis, F.; Roumpos, C. Correlation of the water quality of Plio- Pleistocene aquifers of the greater hydrogeological basin of Ptolemais with the mineralogy and geochemistry of the sediments and the surrounding rocks. In Proceedings of the 11th Hydrogeological Conf. of Greece, Athens, Greece, 4–6 October 2017. (In Greek).
16. Technical Chamber of Greece. *Estimation of the Transition Cost to a Low Lignite Production Economy for the Region. of Western Macedonia*; Technical Chamber of Greece, Department of Western Macedonia: Kozani, Greece, 2012.
17. Sotiropoulos, D.; Karlopoulos, E.; Dimitriou, A.; Soumelidis, A. *Threats for the Region. of Western Macedonia towards an Abolition of Lignite-Based Electricity Production in Greece by 2028*; Reference Study ordered by the Governor of Western Macedonia, Technical Chamber of Greece, Department of Western Macedonia: Kozani, Greece, January 2020; 51p.

18. Pavloudakis, F.; Roumpos, C.; Karlopoulos, E.; Koukouzas, N. Planning and implementing surface mines reclamation works under the new EU strategy for the transition of lignite intensive regions to a post-mining era. In Proceedings of the 12th International Conference of Mine Closure, Leipzig, Germany, 3–7 September 2018; pp. 139–146.

19. Pavloudakis, F.; Agioutantis, Z. Using environmental permits for boosting the environmental performance of large-scale lignite surface mining activities in Greece. In Proceedings of the 2008 National Meeting of the American Society of Mining and Reclamation, Richmond, VA, USA, 14–19 June 2008.

20. Pavloudakis, F.; Sachanidis, C.; Roumpos, C. Planning land reclamation and uses at Ptolemais lignite mines complex. In Proceedings of the 5th Intl. Conference—COAL 2011, Zlatibor, Serbia, 19–22 October 2011.

21. PPC. *Environmental Impact Assessment Study of Ptolemais Lignite Mine*; Public Power Corporation SA: Athens, Greece, 2011.

22. Pavloudakis, F.; Roumpos, C.; Karlopoulos, E.; Koukouzas, N.; Sachanidis, C.G.; Pyrtses, S. Economic and social impact assessment of decarbonisation in Ptolemais lignite mining area. In Proceedings of the 9th International Conference COAL 2019, Zlatibor, Serbia, 23–26 October 2019; pp. 195–208.

23. Profitou–Athanasiadou, D.; Panagiotou, E.; Misopolinos, N. *Agricultural and Techno-Economic Study for the Utilization of Restored Land of Ptolemais and Amynteon Surface Lignite Mines*; Aristotle University of Thessaloniki—School of Veterinary: Thessaloniki, Greece, 2010; 173p.

24. Papadopoulou, F.; Tentsoglidou, M.; Pavloudakis, F.; Papadimopoulos, N.; Papadopoulos, I. Evaluation of Honey Producing Potential of Robinia Pseudacacia in Reforested Old Lignite Mines in West Macedonia. *J. Environ. Sci. Eng.* **2018**, *B7*, 354–359. [CrossRef]

25. Lima, A.T.; Mitchell, K.; O'Connell, D.W.; Verhoeven, J.; Van Cappellen, P. The legacy of surface mining: Remediation, restoration, reclamation and rehabilitation. *Environ. Sci. Policy* **2016**, *66*, 227–233. [CrossRef]

26. Land Rehabilitation Society of Southern Africa. *Land Rehabilitation Guidelines for Surface Coal Mines*; Land Rehabilitation Society of Southern Africa, Coaltech, Minerals Council of South Africa; Coaltech Research Association: Johannesburg, South Africa, 2018.

27. Australian Government. *Mines Rehabilitation—Leading Practice Sustainable Development Program. for the Mining Industry*; Department of Industry, Science, Energy and Resources; Australian Government: Canberra, Australia, September 2016; 76p.

28. Australian Government. *Mines Closure and Completion—Leading Practice Sustainable Development Program. for the Mining Industry*; Department of Industry, Science, Energy and Resources; Australian Government: Canberra, Australia, October 2006; 73p.

29. Australian Government. *Community Engagement and Development—Leading Practice Sustainable Development Program. for the Mining Industry*; Department of Industry, Science, Energy and Resources; Australian Government: Canberra, Australia, September 2016; 72p.

30. Limpitlaw, D.; Aken, M.; Lodewijks, H.; Viljoen, J. Post-mining rehabilitation, land use and pollution at collieries in South Africa. In Proceedings of the Colloquium: Sustainable Development in the Life of Coal Mining, Boksburg, South Africa, 13 July 2005; The Southern African Institute of Mining and Metallurgy: Johannesburg, South Africa, 2005.

31. Pavloudakis, F.; Roumpos, C.; Spanidis, P.-M. Optimisation of resource efficiency and material use in surface mining projects based on a circular economy model. In *Circular Economy and Sustainability*; Elsevier Publications: Amsterdam, The Netherlands, In Press.

32. Food and Agriculture Organisation of the United Nations. *Guidelines for Land Use Planning*; Food and Agriculture Organisation of the United Nations: Rome, Italy, 1993; 135p.

33. Kivinen, S. Sustainable Post-Mining Land Use: Are Closed Metal Mines Abandoned or Re-used Space? *Sustainability* **2017**, *9*, 1705. [CrossRef]

34. European Commission. *European Green Deal Investment Plan, COM (2020) 21 Final*; European Commission: Brussels, Belgium, January 2020.

35. Yuill, C.; Gorton, W.T.; Frakes, M. *A Land Use Decision Methodology for Mine Lands in Appalachia*; The Appalachian Regional Commission: Washington, DC, USA, 1981; 308p.

36. Ostojic, I.; Glažar, T. *Criteria for Evaluation and Guidelines for Land Use Planning*; Igra ustvarjalnosti— The Creativity Game, Scientific Journal no.2; Faculty of Architecture and Faculty of Civil and Geodetic Engineering, University of Ljubljana: Ljubljana, Slovenia, 2014; pp. 24–32. [CrossRef]

37. Mborah, C.; Bansah, K.J.; Boateng, M.K. Evaluating Alternate Post-Mining Land-Uses: A Review. *Environ. Pollut.* **2016**, *5*, 14–22. [CrossRef]

38. Skousen, J.; Zipper, C.E. Post-mining policies and practices in the Eastern USA coal region. *Int. J. Coal Sci. Technol.* **2014**, *1*, 135–151. [CrossRef]

39. Pavloudakis, F.; Galetakis, M.; Roumpos, C. A spatial decision support system for the optimal environmental reclamation of open-pit coal-mines in Greece. *Intl. J. Min. Reclam. Environ.* **2009**, *23*, 291–303. [CrossRef]

40. Palogos, I.; Galetakis, M.; Roumpos, C.; Pavloudakis, F. Selection of optimal land uses for the reclamation of surface mines by using evolutionary algorithms. *Intl. J. Min. Sci. Technol.* **2017**, *27*, 491–498. [CrossRef]

41. Wang, J.; Zhao, F.; Yang, J.; Li, X. Mining Site Reclamation Planning Based on Land Sustainability Analysis and Ecosystem Services Evaluation: A Case Study in Liaoning Province, China. *Sustainability* **2017**, *9*, 890. [CrossRef]

Article

Numerical Analysis of Longwall Gate-Entry Stability under Weak Geological Condition: A Case Study of an Indonesian Coal Mine

Takashi Sasaoka [1], Pisith Mao [1,2,*], Hideki Shimada [1], Akihiro Hamanaka [1] and Jiro Oya [3]

[1] Department of Earth Resources Engineering, Faculty of Engineering, Kyushu University, Fukuoka 819-0395, Japan; sasaoka@mine.kyushu-u.ac.jp (T.S.); shimada@mine.kyushu-u.ac.jp (H.S.); hamanaka@mine.kyushu-u.ac.jp (A.H.)

[2] Department of Geo-Resources and Geotechnical Engineering, Institute of Technology of Cambodia, Phnom Penh 12150, Cambodia

[3] MM Nagata Coal Tech Co., Ltd., Tokyo 140-0002, Japan; j-oya@mitsui-matsushima.co.jp

* Correspondence: mao17r@mine.kyushu-u.ac.jp

Received: 28 July 2020; Accepted: 3 September 2020; Published: 10 September 2020

Abstract: The present research primarily focuses on the investigation of gate-entry stability of longwall trial panel under weak geological condition in Indonesia coal mine by means of numerical analysis. This work aims at identifying appropriate roof support at 100 m and 150 m of depth during gate development. Due to depth depending competency of dominant rock, the stability of gate-entry at 100 m of depth can be optimized by leaving at least 1 m of remaining coal thickness (RCT) above and below the gate-entry. The appropriate support for the trial panel gate-entry is steel arch SS540 with 1 m and 0.5 m spacing for 100 m and 150 m of depth, respectively. The influence of panel excavation on gate-entry is also discussed. Regarding the aforementioned influence, the utilization of additional gate mobile support is recommended at least 10 m from the longwall face.

Keywords: longwall mining; weak geological condition; gate-entry stability; remaining coal thickness

---

1. Introduction

The majority of coal resources in Indonesia is situated in weak geological condition region [1–3]. This weak geological condition presents a big challenge in developing underground longwall coal mines. Since the beginning, the development of underground longwall coal mine in Indonesia has adopted a general longwall guideline developed from other major coal producing countries including the U.S., Australia, and China. As a result, the utilization of bolt-type support is very common for gate-entry support in Indonesia. The instability of gate-entry has become one of the major ground issues for longwall mine development in Indonesia. This problem is mainly caused by inadequate support design and lack of understanding of ground behavior during the longwall mining development. The weak rock usually refers to a group of rock whose uniaxial compressive strength falls below 25MPa [4,5]. This property causes the weak rock to behave totally different from hard rock corresponding to longwall mining operation in various aspects including roof caving behavior, subsidence, abutment stress distribution [6–9]. For this particular research, the stability of gate-entry is linked directly to the behavior of the abutment stress distribution. In general, for all underground opening structures, there are three areas surrounding that opening [8,9]. These include stress decrease area, which is located above the opening, stress increase area, which contains the highest peak stress at a certain distance from the opening, and the initial stress area, which marks the end of stress disturbance from the opening. The softer the rock measure, the farther the peak stress is located from the opening. This could cause a problem in longwall mining, as the abutment stress link directly to the stability of the gate-entry. Weak rock also reduces the efficiency of active support such as tensioned rockbolts and cablebolts, which are the most common support systems in longwall mining while favoring passive support such as steel arch and shotcrete [10–14].

In longwall mining, gate-entries provide access for mining equipment transportation to the longwall face and carried out coal production from the longwall face to the surface. Gate-entries also work as a major component of longwall ventilation system to remove the toxic gases as well as dust particles released during panel extraction [15]. Thus, the stability of gate-entry is really crucial to ensure a smooth longwall mining operation. Two major aspects influence the stability of gate-entry [16]. First is during the excavation operation of the gate-entry itself due to stress redistribution around the gate opening. Second is during longwall panel extraction. During this operation, a large portion of coal is excavated, which causes the roof above to fracture and fall down. This fractured zone is called the goaf area. This operation results in a huge stress redistribution to occur in the surrounding area. As shown in Figure 1, due to the development of fracture zone in the goaf area, overburden stress is redistributed into the rib side around the goaf area. This phenomenon causes high abutment stress along longwall face and peak stress at both edges of the longwall face [17]. Furthermore, this phenomenon affects the stability of the gate-entry as both gate-entry of the longwall is connected to the edge of the longwall face where the peak stress is located.

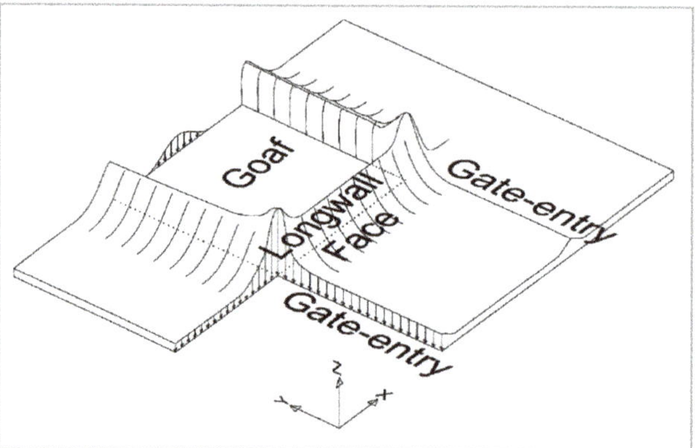

Figure 1. Vertical Stress Distribution during Longwall Panel Extraction.

There are multiple factors that limit mining height in the longwall mine design, in which the weak geological condition is also included [18]. In the case where the thickness of the coal seam is significantly thicker than the mining height, there must be some portion of coal left above and below the excavation area. This remained portion of coal is quite advantageous for improving the stability of the longwall mine, especially when the surrounding rock is relatively weaker than the remained coal. There are abundant research works that study on longwall mining development in thick coal seam [18–21]. Some also integrate weak geological conditions [1,22]. However, none of them have discussed the influence of remaining coal thickness (RCT) on the stability of the longwall development.

This paper discusses the stability of gate entry under various stages of longwall development with two different depths of 100 m and 150 m. This research involves investigating and modeling a trial panel from the actual mine site in Indonesia to evaluate the appropriate support system for the gate-entry. The first part of the paper focuses on the influence of remaining coal thickness on improving the stability of gate-entry. This part is targeted only at a depth of 100 m, as claystone in this area is relatively weaker than coal. The second part of this research is to evaluate the appropriate support system for the trial panel during gate-entry development for 100 m and 150 m of depth. The last section of this research work is to investigate the gate-entry stability under the influence of longwall panel extraction. This operation causes the increment of the stress surrounding gate-entry, which is located near the longwall face. This stress might exceed the maximum support capacity of the designed support, which results in instability of gate-entry. Additional mobile support is necessary for maintaining the stability of the gate-entry during panel extraction. The length of the additional support along the gate is also evaluated in this section.

2. General Overview of the Targeted Indonesia Coal Mine

The research is conducted in a coal mine, which is situated in the area of Kutai Kertanegara of East Kalimantan Island. The analytical of drill hole core samples revealed that the study area has a monocline structure with claystone as the dominant rock. The dip of the coal seam range between 3° and 13° and the thickness of the coal seam can reach up to almost 10 m in a certain seam. This mine selected the longwall mining method for coal recovery.

The development of longwall mining is sited in a weak geological area. The competence of the dominant rock claystone is related to the depth in which the shallower depth zone has relatively lower strength. The recent development of trial panel which extend between 100 m to 150 m of depth has been gone under operation. Figure 2 shows the top view of the trial panel layout. The dimension of this trial panel is 55 m width by 200 m length with 3 m height. Extensometer (EX) and telltale (TT) are installed in the roof of the gate in order to monitor the deformation of the gate roof.

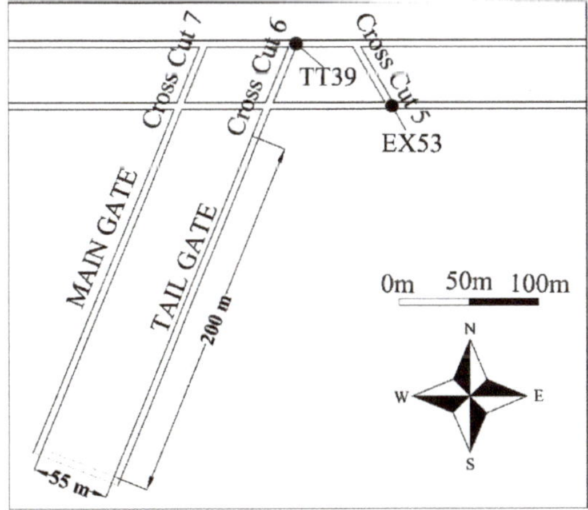

Figure 2. Trial panel and location map of displacement monitoring devices (TT 39 and EX53).

3. Modeling

3.1. General Model Description

The trial panel is modeled in FLAC3D 5.00 software (Fast Lagrangian Analysis of Continua in 3 Dimensions) from Itasca Consulting Group, Inc. Only half of the trial panel, along the panel length, is modeled to save running time. Mohr-Coulomb constitutive model is adopted for all numerical models in this research. Due to the natural incline of the coal seam, the depth of this trial panel ranges from 100 m to 150 m. Both depths are adopted as case studies. General dimensions of the model can be described as 153 m × 60 m × 300 m for 100 m of depth and 203 m × 60 m × 300 m for 150 m of depth as shown in Figure 3. Mining height is 3 m of the coal seam, which is surrounded by dominant rock claystone as the roof and floor. Figure 4 shows the installation of the steel arch in the model. Beam structural elements are used to model steel arch for roof support along the gate-entry. According to the FLAC3D manual of Itasca_Consulting_Group [23], beam structure element is a straight two-noded, finite elements with six degrees of freedom per node. Beam structure behaves as a linearly elastic structure with no failure limit. Each steel arch is modeled from a collection of beams.

Since the steel arch are modeled to attach to the inner part of the excavated gate in such a way that the movement or deformation of the grid will generate force and bending moments develop within the beam structure.

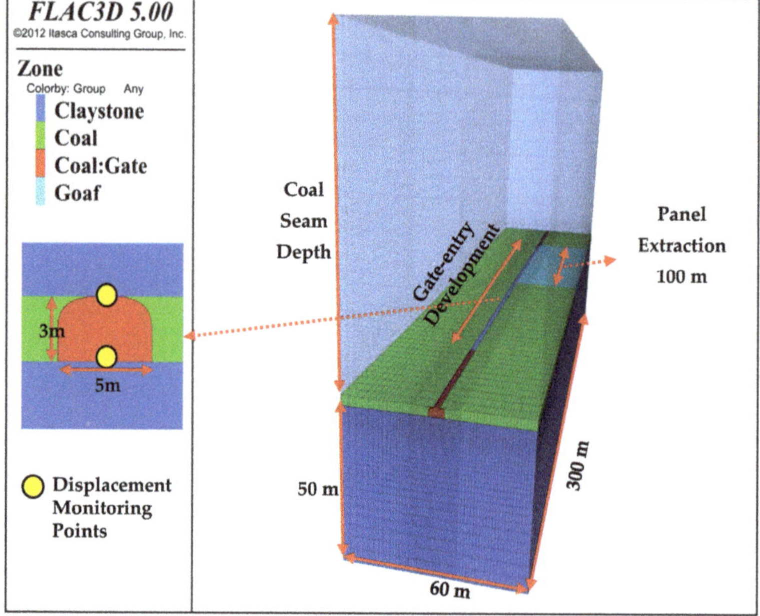

Figure 3. Model Dimensions.

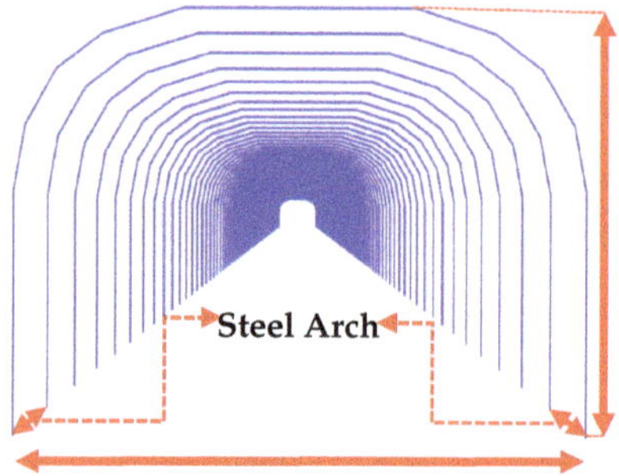

Figure 4. Modelling of steel arch for roof support.

As aforementioned, the result obtained from a series of laboratory tests show that the competence of claystone is related to the depth where relatively lower strength rock can be found in shallower depth. As a result, properties of the claystone are separated into two categories based on the depth. The rock mechanical properties are listed in Table 1. Steel support, composed of three main parts jointed together by eight hex bolts, is utilized for supporting gate-entry. The material type of steel SS540, which has a yield strength of 540 MPa, is selected as the material for the steel arch support. Its mechanical properties are listed in Table 2. Case studies for investigating the influence of the remaining coal thickness are shown in Figure 5. The thickness of the remaining coal ranges from 0 m to 2.5 m above and below the gate. A series of parametric studies are conducted to discuss the appropriate support design under different conditions such as stress ratio K: 1, 1.5, and 2, mining depth: 100 m and 150 m, support spacing: 0.5 m and 1 m. The numerical modeling steps includes the following:

- Construct model geometry (half-width of the panel is modeled to reduce the model analyzing time).
- Apply boundary condition, assign mechanical properties, determine initial stress condition, and analyze for the initial state.
- Excavate gate-entry and install steel arch 200 m.
- Monitor maximum axial stress of steel arch and roof displacement
- Excavate longwall panel for 100 m.
- Monitor maximum support axial stress

Table 1. Mechanical properties of the rock.

Rock	Claystone (100 m of Depth)	Claystone (150 m of Depth)	Coal Seam	Goaf Area
Uniaxial Compressive Strength (MPa)	4.84	6.96	8.16	-
Density (kg/m^3)	2108	2121	1380	1700
E (MPa)	805	1344	1300	15
υ	0.28	0.28	0.32	0.25
C (MPa)	0.6	0.84	2.63	0.001
φ (°)	37.5	41.8	45.6	25
σ_T (MPa)	0.52	0.66	0.58	-

Table 2. Roof support mechanical properties.

Density (kg/m^3)	E (GPa)	υ	Cross Area (cm^2)	Yield (MPa)	Iy ($\times10^{-8}$ m^4)	Iz ($\times10^{-8}$ m^4)	J ($\times10^{-8}$ m^4)
7800	200	0.3	36.51	540	732	154	22

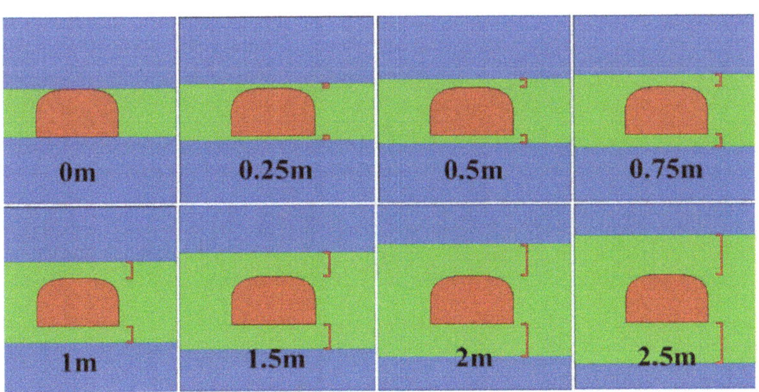

Figure 5. Study Case of Remaining Coal Thickness.

3.2. Goaf Modelling

Goaf is the fragmentation of the immediate roof behind the longwall face caused by the advancing of longwall excavation [24]. The most abutment stress in the excavation area is carried by the pillar. However, a small amount of abutment stress will remain to be carried by goaf [25].

There are several methods to simulate the behavior of goaf. According to Song and Chugh [26], goaf can be simulated using goaf loading characteristics. However, the challenge of quantitative measurements of goaf loading is high due to the inaccessibility in the goaf area. Many assumptions have to be made for obtaining goaf loading characteristics. It is also possible to simulate the longwall panel retreat without considering goaf [27]. The simulation consisted of removing coal from the panel without filling back any material. This space of the null zone material is kept during the analysis of the model. However, this method is applicable only for a small null space. When this null space becomes larger and larger during the advancing of the longwall face, it is possible that the unbalance force in the model is converged into none-zero value. According to the FLAC3D manual of Itasca_Consulting_Group [23], this indicates failure and plastic flows occur within the model. Another possibility when adopting this method is the divergence of the unbalance force, which will affect the accuracy of the analysis.

There is another way of modeling goaf. After panel excavation, a softer material is used to replace both the panel and the caved zone above the excavation area to simulate the collapse of the immediate roof [3,28,29]. The caved zone height can be assessed by using Equation (1):

$$H_C = \frac{100h}{c_1 h + c_2} \tag{1}$$

where H_c represents the height of the caved zone, h is mining height, and c_1 and c_2 are coefficients depending on lithology. Table 3 presents the value of the lithology coefficient depending on rock competence. This paper adopts this method for modeling goaf. According to this equation, the height of the caved zone can be estimated to be 5.93 m. Figure 6 shows the schematic view of the longwall face and goaf during panel extraction.

Table 3. Lithology coefficients.

Lithology	Uniaxial Compressive Strength (MPa)	c_1	c_2
Strong	>40	2.1	16
Medium	20–40	4.7	19
Weak	>20	6.2	32

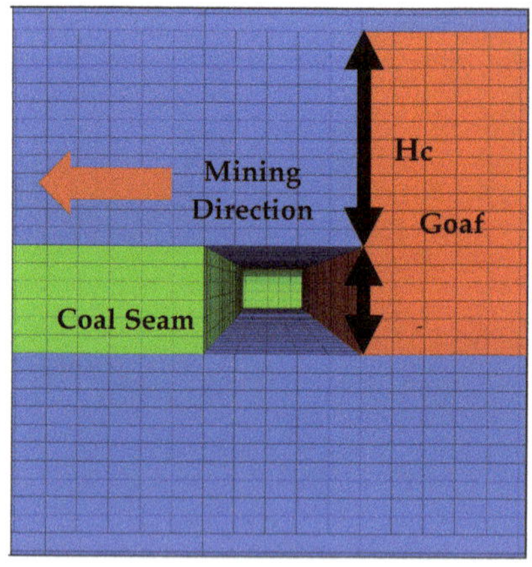

Figure 6. Goaf modeling during panel extraction.

3.3. Modeling for Shield Support

There is a couple of ways of simulating shield support specifically for FLAC3D. The first method is using beam structural elements to model the shield support [30]. However, it is really complex when it comes to assigning the properties of the beam to simulate different setting pressures of shield support. Song et al. [31] came up with a simple and effective method of applying pressure to the roof and the floor of the longwall face to simulate different setting pressure of the shield support. The second method is adopted to simulate shield support for this research.

Due to the fact that this trial panel is in the early development stage, the type and capacity of the shield support for the longwall face are still being considered. For this reason, a random ZY6800/16/32 is selected to support the longwall face for this study. According to Song et al. [31], these shields are equipped with 6800 kN or 23.6 MPa loading capacity, and shield height range from 1.6 m to 3.2 m. These shield support legs are attached at the two-thirds length of the canopy from the tip. As a result, stress is concentrated at the end part of the canopy. For footplates, on the other hand, the leg position is located at the center length of the plate, which generates constant stress distribution at the base of the canopy. Figure 7 shows the development of shield support distribution pressure used in this research for simulations, which is based on the actual distribution pressure profile of ZY6800/16/32. Shield setting pressure is selected to be 6000 kN.

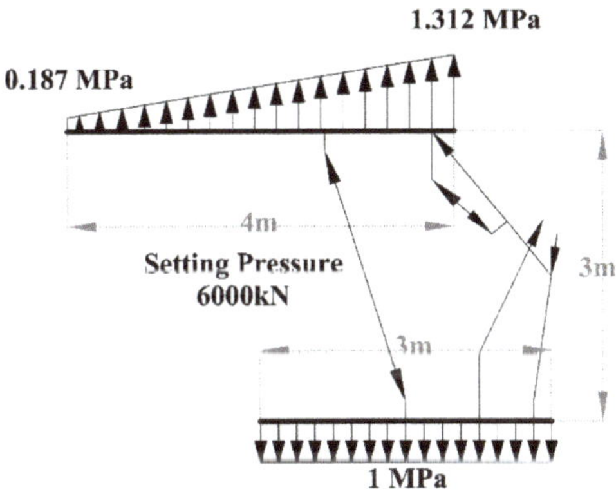

Figure 7. Pressure distribution for modeling shield support.

3.4. Model Validation

The simulation result is validated with the field measurement results for displacement. Two types of roof displacement monitoring devices are installed along the main roadway, which include a telltale (TT) and extensometer (EX). TT39 and EX53 are used for this validation; the installation location can be found in Figure 2. The installation location is at the T-section area, around 100 m from the surface. The rock surrounding the gate is claystone. One meter of support spacing is used to support this gate. These devices are installed after gate development operation by drilling a small vertical borehole into the roof of the gate around 6 m in length and anchoring these monitoring devices along the borehole. These devices can monitor gate roof movement in this 6 m installed range. Gate roof movement and displacement has been recorded daily in millimeter. The model is constructed in FLAC3D based on the site described above.

The displacement result from the simulation model is used to compare with the results of the extensometer and telltale from the field. Figure 8 shows the comparison of the results between the telltale and simulation model. The total displacement of the simulation result is 13.3 mm. This value is barely distinguishable from the total displacement of telltale, which is 14 mm. The comparison between the displacement results from the simulation model and the extensometer in Figure 9 also shows a great agreement. The maximum roof displacement from the extensometer is 12.6 mm, which is less than a millimeter difference from the simulated result. The result of the extensometer seems to be still increasing and the measured value may exceed the computed value within a certain time. However, this might not be a problem for estimating gate-entry stability during panel operation. The gate-entry is a temporary structure that is expected to collapse at the end of each panel extraction operation. The field monitoring time is more than 200 days. As a result, this period is more than enough to finish the panel extraction operation for a typical panel. The overall comparisons indicate that this model configuration is suitable for predicting rock behavior in this research.

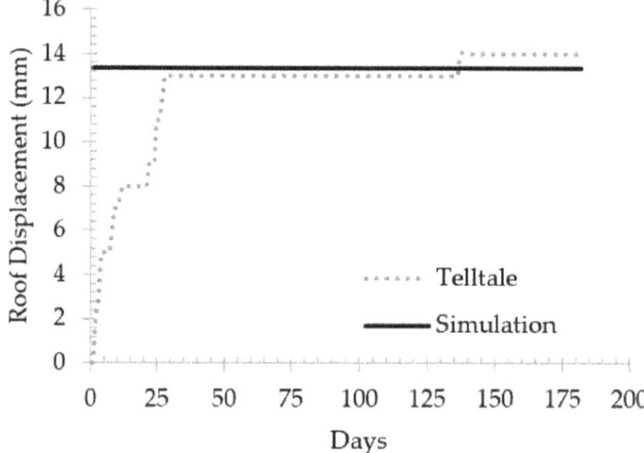

Figure 8. Comparison between telltale results and simulation results.

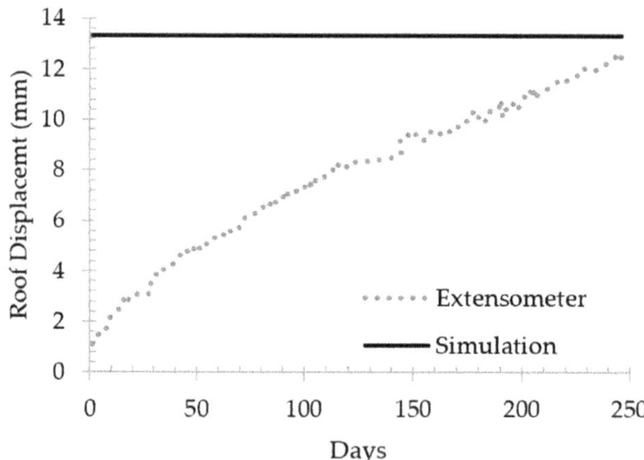

Figure 9. Comparison between extensometer results and simulation results.

4. Results and Discussions

4.1. Stability of Gate-Entry under the Influence of the Remaining Coal Thickness (RCT)

First, the simulation models are constructed to assess the influence of the remaining coal thickness (RCT) on improving gate-entry stability at a depth of 100 m. In order to analyze the pure influence of RCT, gate entry is excavated without any support in the model. After gate excavation, displacement is monitored at the middle point of the roof and floor of the gate-entry as illustrated in Figure 3. The results are shown in Figures 10 and 11. As predicted, the displacement decreases with the increase of RCT.

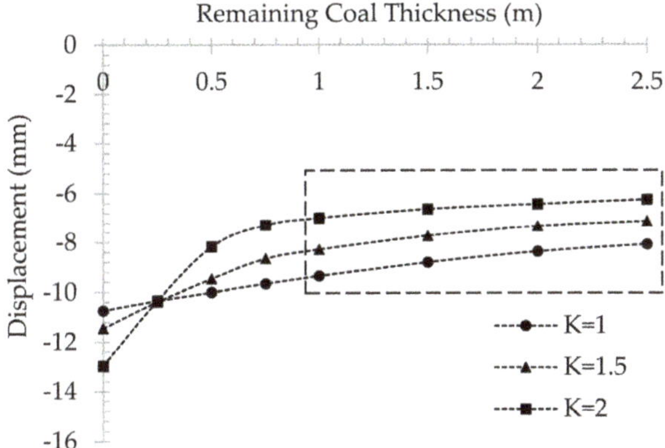

Figure 10. Correlation between roof displacements and RCT.

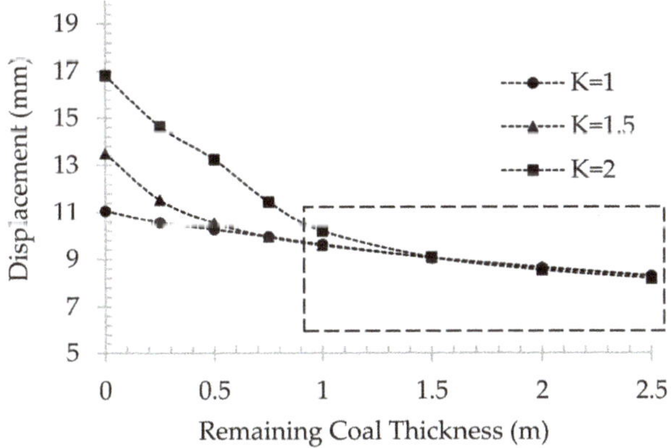

Figure 11. Correlation between floor displacements and RCT.

Figure 10 shows the correlation between roof displacements and remaining coal thickness with different stress ratios, while Figure 11 shows the correlation between floor displacement and remaining coal thickness. The reductions of displacement on the roof are around 2.5 mm, 4.5 mm, and 6.5 mm for stress ratios 1, 1.5, and 2, respectively. For the floor, reductions value for the respective stress ratios are 2.5 mm, 5 mm, and 8.5 mm. It is quite difficult to draw any conclusion from the displacement result obtained from stress ratio 1 due to the fact that the result trend line decreased in a form of constant linear reduction when RCT increased. For bigger stress ratios of 1.5 and 2, both roof and floor results show a noticeably huge decrease of displacement when RCT is between 0 m to 1 m. Above 1 m of RCT, these shrinkages of displacement are very minimal. From these results, it can be said that 1 m of RCT is the optimum thickness for improving gate-entry stability.

4.2. Gate-Entry Stability during Gate Development

The stability of the gate-entry is maintained by a steel arch support. The results of support axial stress for each case are used for comparing with the support maximum yield strength. If the support axial stress is greater than support yield strength, gate-entry is considered to be in the unstable condition. A wide support spacing of 1 m is adopted for gate-entry at 100 m of depth. According to the previous section, 1 m of RCT is the optimum thickness. The incorporation of RCT influence is only assessed for cases at 100 m of depth. For 150 m of depth, as the competence of coal and claystone are quite similar, RCT does not have any influence on gate-entry. There are two spacings of steel support: 1 m and 0.5 m, which are adopted for 150 m of depth due to the higher stress. The result of the maximum support axial stress of 1 m of RCT and without RCT for 100 m of depth are illustrated in Figure 12. The outcome indicated that for 100 m of depth, if 1 m of RCT is available, the utilization of steel arch SS540 with 1 m spacing to support gate-entry is applicable up to K = 1.9. On the other hand, if RCT is not available, this support is only suitable up to K = 1.6. From this result, we are able to identify the potential reduction of support cost for stress ratios between 1.6 and 1.9. Within this stress ratio range, the existence of RCT can hypothetically enlarge the required support spacing from 0.5 m to 1 m, which translates to a reduction of the support cost by half. Maximum support axial stress for 150 m of depth is shown in Figure 13. According to this figure, 1 m support spacing is capable to maintain the stability of the gate-entry up to K = 1.1, while a narrower spacing of 0.5 m can increase the support capacity to withstand K = 1.6. Higher support capacity is required if the stress ratio is above 1.6, which can be achieved by narrowing the support spacing or incorporate the current steel arch support with a rock bolt or shotcrete.

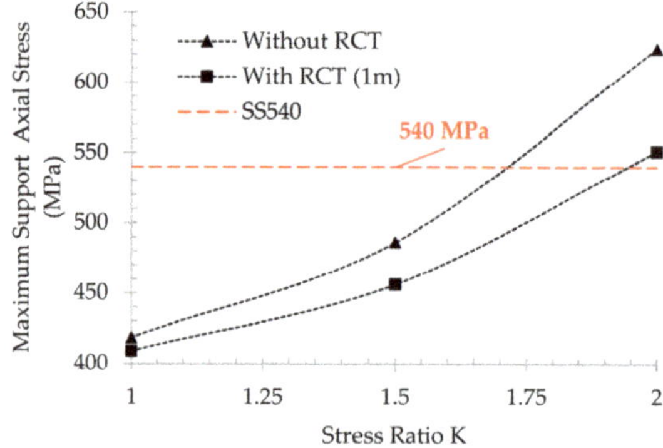

Figure 12. Maximum support axial stress for 100 m of depth (1 m spacing).

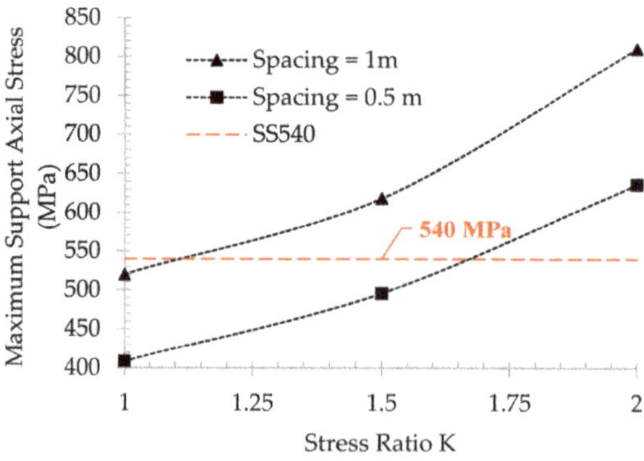

Figure 13. Maximum support axial stress for 150 m of depth.

4.3. Gate-Entry Stability during Panel Extraction

The impact of panel extraction on the stability of gate-entry is crucial in longwall mining study. Generally, the stress increases significantly in some distance along the entry-gate from the longwall face. These over-stress distances are important for optimizing additional mobile support during panel extraction operation.

4.3.1. Vertical Stress Distribution

After advancing the longwall face for 100 m, the vertical stress is monitored above the mining panel surround the longwall face. Three-dimensional (3D) schematic plots of vertical stress distributions are shown in Figures 14 and 15 for 100 m and 150 m, respectively. A similar stress distribution can be observed from both schematics, which include stress release in the goaf area, high-stress concentration in the longwall face, and higher peak stress in the corner between the longwall face and gate-entry. Figures 16 and 17 show the vertical stress along the transverse distance, which is monitored from the longwall face range from 0 m to 10 m with an interval of 2 m. The location of the monitoring lines can be found in Figures 14 and 15.

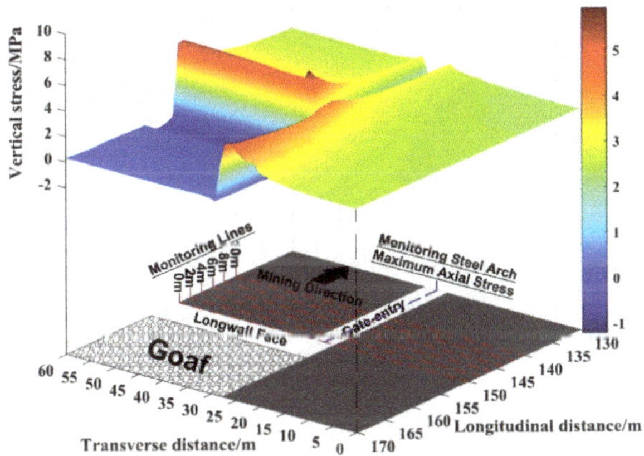

Figure 14. Three-dimensional (3D) schematic of the vertical stress above the mining panel at 100 m of depth.

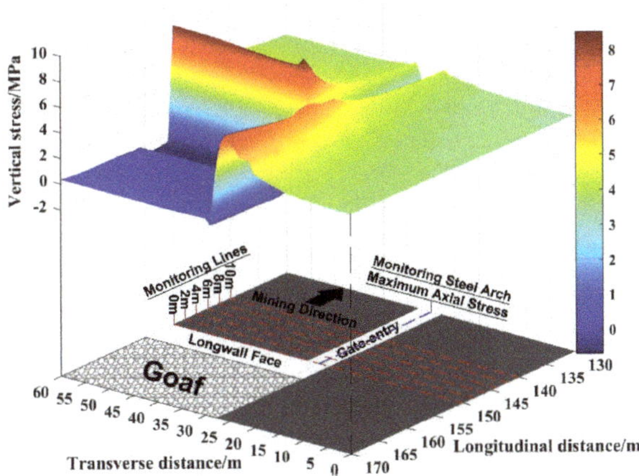

Figure 15. 3D schematic of the vertical stress above the mining panel at 150 m of depth.

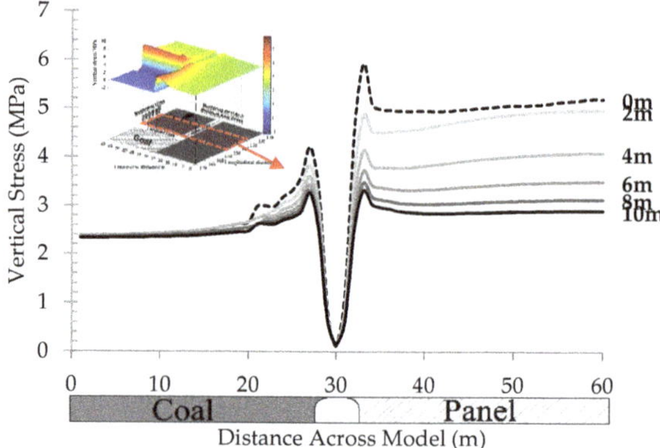

Figure 16. Vertical stress with different distance from the longwall face for 100 m of depth.

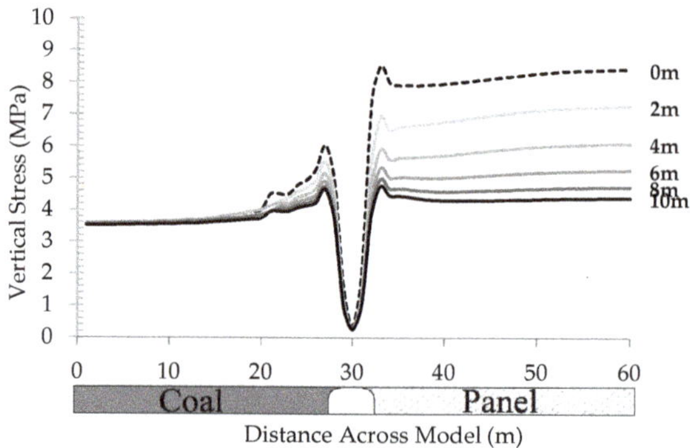

Figure 17. Vertical stress with different distance from the longwall face for 150 m of depth.

Figure 16 shows vertical stress result in case of 100 m of depth. The observation peak stress of both sides of the gate near to the longwall face is asymmetrical, which is caused by the stress redistribution due to the panel extraction. The highest peak stress at 0 m from longwall face is around 6 MPa on the panel side and 4 MPa on the coal side. When the monitoring line is moved away from the coal face the peak stress for both sides of the gate tend to become more and more symmetrical, which indicates that the influence of the panel extraction operation decrease. Around 10 m distance from the longwall face, the peak of vertical stress becomes almost symmetrical with the peak stress around 3 MPa for both sides of the gate.

The same statement is also true for Figure 17 in case of 150 m of depth. At 0 m from the longwall face, the highest peak of vertical stress is around 8.5 MPa on the panel side and 6 MPa on the coal side. For 10 m distance from the longwall face, the peak stress reduces to 4.5 MPa for both sides of the gate. This nearly symmetrical peak of vertical stress also designates less influence of panel extraction on gate-entry. These results point to the likelihood that the influence of panel extraction is at their highest impact in the longwall face. This influence tends to decrease with the increase of distance from the longwall face. Based on these results of vertical stress analysis, the influence of panel extraction decreases to their minimum from 10 m distance from the longwall face.

4.3.2. Maximum Axial Stress of Each Individual Support along Gate-Entry

The maximum axial stress of each individual steel arch support is a great indicator of the impact of panel extraction on the stability of gate-entry. As previously explained, in the case where support axial stress exceeds the maximum support yield strength, instability of the gate will occur. For this reason, the value of maximum axial stress of each individual steel arch used to support gate-entry is recorded from the longwall face to the direction following the mining direction. Figures 14 and 15 also show the location of maximum axial stress monitoring.

Figures 18 and 19 show maximum support axial stress along the gate-entry for 100 m and 150 m of depth, respectively. The results indicated that the maximum axial stress along gate-entry increased significantly for the distance close to the longwall face. The distance of over-stress can be obtained by comparing this maximum axial stress results to the yield strength of SS540. These results show that the over-stress distance of both 100 m of depth with 1 m spacing and 150 m of depth with 0.5 m spacing is approximately 10 m. It is necessary to increase the capacity of the support system by installing mobile support in this over-stress distance. The mobile support should be able to move along when the longwall face advances. Steel arch support with 1 m spacing can be used to support the gate-entry for 150 m of depth during gate excavation. However, as seen in Figure 19, the over-stress distance for 1 m spacing case exceeds 80 m in length. The longer the length, the bigger amount of mobile support that is required to maintain the stability of gate-entry. As a result, it is suggested to select a narrower 0.5 m spacing for maintaining the stability of gate-entry for 150 m of depth.

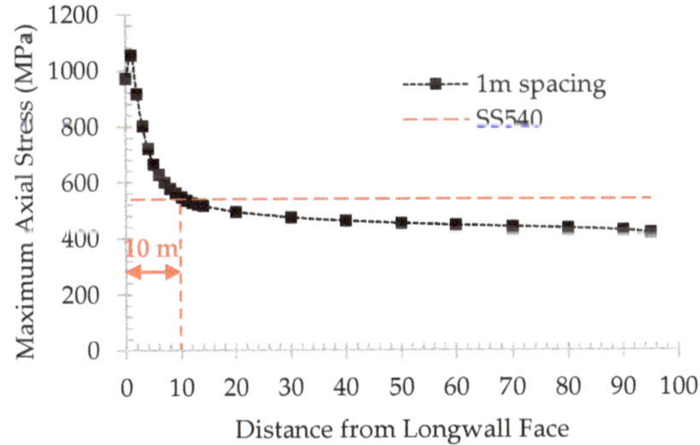

Figure 18. Maximum support axial stress along the gate-entry for 100 m of depth.

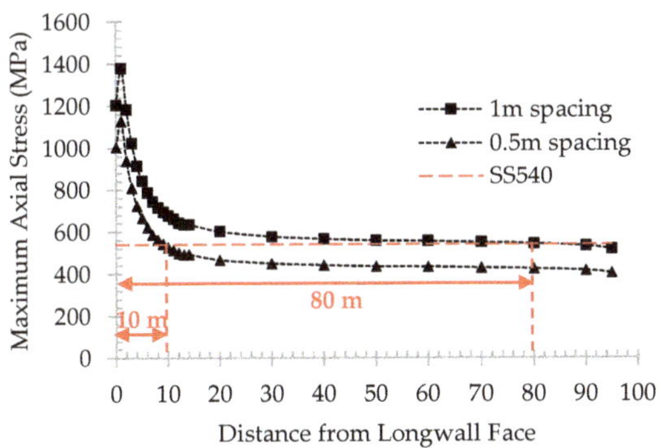

Figure 19. Maximum support axial stress along the gate-entry for 150 m of depth.

5. Conclusions

The investigation of gate-entry stability of the underground longwall mining under weak geological conditions has been carried out by numerical modeling to evaluate the most appropriate support system. The influence of remaining coal thickness has also been assessed by observing the roof sag and floor heave of the gate-entry. The gate-entry stability has been inspected at both stages of the gate development and panel extraction by monitoring the support axial stress. The summary of this research can be described as the following:

(1) In the shallow depth where the rock is relatively weaker, the optimum thickness of 1 m of RCT can help to improve gate-entry stability by extending the overall support capacity, which is suitable for stress ratio 1.6 when adopting 1 m spacing, to withstand a higher stress ratio up to 1.9 when using the same spacing configuration.

(2) The appropriate support configuration for 150 m of depth is 0.5 m spacing where the stress ratio is 1.6. A narrower spacing or incorporation of rock bolt or shotcrete is required in the area where the stress ratio is higher than 1.6.

(3) Results from the vertical stress are in line with the results of maximum support axial stress. During panel excavation, for both cases, that is, 100 m of depth with 1 m spacing and 150 m of depth with 0.5 m spacing support configuration, the unstable distance along gate-entry is approximately 10 m from the longwall face.

To sum up, a steel arch is an effective support system for gate-entry of underground longwall mining under weak geological condition. The availability of RCT helps improving the stability of the gate-entry especially in the shallow depth. The narrower support spacing is required for deeper longwall mining depth. In addition, due to the influence of panel extraction, higher stress concentration can be seen in the surrounding area of the longwall face. As a result, the end portion of gate-entry, which is connected to the longwall face, becomes unstable. To solve this problem, the installation of additional mobile support, which can be moved according to the retreat operation of the longwall face, is required in that unstable portion of gate-entry.

Author Contributions: Conceptualization, T.S. and P.M.; methodology, T.S. and P.M.; software, P.M.; validation, T.S., H.S., A.H., J.O. and P.M.; formal analysis, T.S. and P.M.; investigation, T.S. and P.M.; resources, J.O.; data curation, T.S. and P.M.; writing—original draft preparation, P.M.; writing—review and editing, T.S., H.S., A.H. and P.M.; visualization, P.M.; supervision, T.S., H.S., A.H. and J.O. All authors have read and agreed to the published version of the manuscript.

Funding: This research received no external funding.

Conflicts of Interest: The authors declare no conflict of interest.

References

1. Sasaoka, T.; Hamanaka, A.; Shimada, H.; Matsui, K.; Lin, N.Z.; Sulistianto, B. Punch multi-slice longwall mining system for thick coal seam under weak geological conditions. *J. Geol. Resour. Eng.* **2015**, *1*, 28–36.

2. Sasaoka, T.; Karian, T.; Hamanaka, A.; Shimada, H.; Matsui, K. Application of highwall mining system in weak geological condition. *Int. J. Coal Sci. Technol.* **2016**, *3*, 311–321. [CrossRef]

3. Pongpanya, P.; Sasaoka, T.; Shimada, H.; Hamanaka, A.; Wahyudi, S. Numerical Study on Effect of Longwall Mining on Stability of Main Roadway under Weak Ground Conditions in Indonesia. *J. Geol. Resour. Eng.* **2017**, *3*, 93–104.

4. Hoek, E.; Brown, E.T. Practical estimates of rock mass strength. *Int. J. Rock Mech. Min. Sci.* **1997**, *34*, 1165–1186. [CrossRef]

5. Bieniawski, Z.T. Estimating the strength of rock materials. *J. S. Afr. Inst. Min. Metall.* **1974**, *74*, 312–320. [CrossRef]

6. Guney, A.; Gul, M. Analysis of surface subsidence due to longwall mining under weak geological conditions: Turgut basin of Yatağan-Muğla (Turkey) case study. *Int. J. Min. Reclam. Environ.* **2019**, *33*, 445–461. [CrossRef]

7. Cui, X.; Zhao, Y.; Wang, G.; Zhang, B.; Li, C. Calculation of Residual Surface Subsidence Above Abandoned Longwall Coal Mining. *Sustainability* **2020**, *12*, 1528. [CrossRef]

8. Yang, R.; Zhu, Y.; Li, Y.; Li, W.; Lin, H. Coal pillar size design and surrounding rock control techniques in deep longwall entry. *Arab. J. Geosci.* **2020**, *13*, 1–14. [CrossRef]

9. Li, Y.; Lei, M.; Wang, H.; Li, C.; Li, W.; Tao, Y.; Wang, J. Abutment pressure distribution for longwall face mining through abandoned roadways. *Int. J. Min. Sci. Technol.* **2019**, *29*, 59–64. [CrossRef]

10. Hong-Pu, K.; Jian, L.; Yong-Zheng, W. Development of high pretensioned and intensive supporting system and its application in coal mine roadways. *Procedia Earth Planet. Sci.* **2009**, *1*, 479–485. [CrossRef]

11. Jiao, Y.-Y.; Song, L.; Wang, X.-Z.; Coffi Adoko, A. Improvement of the U-shaped steel sets for supporting the roadways in loose thick coal seam. *Int. J. Rock Mech. Min. Sci.* **2013**, *60*, 19–25. [CrossRef]

12. Meng, Q.; Han, L.; Qiao, W.; Lin, D.; Fan, J. Support technology for mine roadways in extreme weakly cemented strata and its application. *Int. J. Min. Sci. Technol.* **2014**, *24*, 157–164. [CrossRef]

13. Phanthoudeth, P. *Appropriate Design of Longwall Coal Mining System under Weak Geological Conditions in Indonesia*; Kyushu University: Fukuoka, Japan, 2018.

14. Toraño, J.; Díez, R.R.G.; Rivas Cid, J.M.; Barciella, M.M.C. FEM modeling of roadways driven in a fractured rock mass under a longwall influence. *Comput. Geotech.* **2002**, *29*, 411–431. [CrossRef]

15. Brune, J.; Sapko, M. A modeling study on longwall tailgate ventilation. In Proceedings of the 14th United States/North American Mine Ventilation Symposium, Salt Lake City, UT, USA, 17–20 June 2012.

16. Jiang, L.; Sainoki, A.; Mitri, H.S.; Ma, N.; Liu, H.; Hao, Z. Influence of fracture-induced weakening on coal mine gateroad stability. *Int. J. Rock Mech. Min. Sci.* **2016**, *88*, 307–317. [CrossRef]

17. Suchowerska, A.M.; Merifield, R.S.; Carter, J.P. Vertical stress changes in multi-seam mining under supercritical longwall panels. *Int. J. Rock Mech. Min. Sci.* **2013**, *61*, 306–320. [CrossRef]

18. Ozfirat, M.; Simsir, F.; Gonen, A. A brief comparison of longwall methods used at mining of thick coal seams. In Proceedings of the 19th International Mining Congress and Fair, Izmir, Turkey, 9–12 June 2005.

19. Hebblewhite, B. Status and prospects of underground thick coal seam mining methods. In Proceedings of the 19th International Mining Congress and Fair, Izmir, Turkey, 9–12 June 2005.

20. Matsui, K.; Sasaoka, T.; Shimada, H.; Furukawa, H.; Takamoto, H.; Ichinose, M. Some considerations in underground mining systems for extra thick coal seams. *Coal Int.* **2011**, *259*, 38–43.

21. Wang, J.C.; Wang, Z.H.; Yang, S.L. Stress analysis of longwall top-coal caving face adjacent to the gob. *Int. J. Min. Reclam. Environ.* **2020**, *34*, 476–497. [CrossRef]

22. Zarlin, N.; Sasaoka, T.; Shimada, H.; Matsui, K. Numerical study on an applicable underground mining method for soft extra-thick coal seams in Thailand. *J. Eng.* **2012**, *4*, 739–745. [CrossRef]

23. Itasca_Consulting_Group. *FLAC3D 5.0 Manual*; Itasca Consulting Group: Minneapolis, Minnesota, 2012.

24. Wang, J.; Yang, S.; Li, Y.; Wang, Z. A dynamic method to determine the supports capacity in longwall coal mining. *Int. J. Min. Reclam. Environ.* **2015**, *29*, 277–288. [CrossRef]

25. Thin, I.G.T.; Pine, R.J.; Trueman, R. Numerical modelling as an aid to the determination of the stress distribution in the goaf due to longwall coal mining. *Int. J. Rock Mech. Min.* **1993**, *30*, 1403–1409. [CrossRef]

26. Song, G.; Chugh, Y. 3D analysis of longwall face stability in thick coal seams. *J. S. Afr. Inst. Min. Metall.* **2018**, *118*, 131–142. [CrossRef]

27. Zhao, T.; Liu, C.; Yetilmezsoy, K.; Gong, P.; Chen, D.; Yi, K. Segmental adjustment of hydraulic support setting load in hard and thick coal wall weakening: A study of numerical simulation and field measurement. *J. Geophys. Eng.* **2018**, *15*, 2481–2491. [CrossRef]

28. Cheng, Y.; Wang, J.; Xie, G.; Wei, W. Three-dimensional analysis of coal barrier pillars in tailgate area adjacent to the fully mechanized top caving mining face. *J. Rock Mech. Min.* **2010**, *47*, 1372–1383. [CrossRef]

29. Yavuz, H. An estimation method for cover pressure re-establishment distance and pressure distribution in the goaf of longwall coal mines. *J. Rock Mech. Min.* **2004**, *41*, 193–205. [CrossRef]

30. Wang, Z.-H.; Yang, J.-H.; Meng, H. Mechanism and controlling technology of rib spalling in mining face with large cutting height passing through fault. *J. China Coal Soc.* **2015**, *40*, 42–49.

31. Song, G.; Chugh, Y.P.; Wang, J. A numerical modelling study of longwall face stability in mining thick coal seams in China. *J. Min. Miner. Eng.* **2017**, *8*, 35–55. [CrossRef]

Article

Comparative Analysis of the Mining Cribs Models Filled with Gangue

Krzysztof Skrzypkowski

Faculty of Mining and Geoengineering, AGH University of Science and Technology, Mickiewicza 30 av., 30-059 Kraków, Poland; skrzypko@agh.edu.pl; Tel.: +481-2617-2160

Received: 8 September 2020; Accepted: 10 October 2020; Published: 12 October 2020

Abstract: In the article, comparative laboratory strength tests of three-point and four-point wooden cribs models are presented. In the case of cribs with a triangular cross-section, the notches made at an angle of 60 degrees were used for the first time. The individual beams of the three-point and four-point cribs were stacked horizontally and connected to each other by means of a quick-drying adhesive. The main aim of the research was to compare the empty models with cribs filled with a gangue. In order to better understand the mechanism of behavior of models under loads, load-displacement and pressure-compressibility characteristics are presented. It was found that filling the three-point and four-point crib with gangue increases its maximal load several times compared to the empty cribs.

Keywords: wooden cribs; gangue; load-displacement characteristics; pressure-compressibility characteristics

1. Introduction

Contemporary underground mining is closely related to environmental protection through the use of waste rocks found in the heaps [1–3]. Rockfill is one of the oldest ways of filling post-exploitation spaces. Currently, this type of filling is widely used in the exploitation of hard coal, ores, and chemical raw materials. Yang et al. [4] stated that gangue backfilling mining technology constitute green mining technologies, especially to solve coal gangue emission. Li and Guo [5] and Sun et al. [6] revealed that filling the goaf area with gangue reduces the moving of rock mass. Cao et al. [7] presented wide strip backfilling method in order to control surface subsidence. Tai et al. [8] brought close mining technology of fully mechanized gangue backfilling with the connection of fully mechanized coal mining.

Filling the post-exploitation spaces with the application of rockfill is usually carried out by means of belt conveyors, self-propelled machines [9], scrapers [10], by gravity, and using compressed air pipelines. In Poland, gangue is used in the Legnica-Głogów Copper District of KGHM (Copper Mining and Smelting Complex) mines, especially for the room and pillar method with roof bending [11,12]. The waste rocks are delivered to the goaf using loading machines. Rockfill is very environmentally friendly, because waste rocks from preparatory and exploitation excavations can be directly used for backfilling, without having to remove them to the surface. In addition, it can be used in unconventional different operating conditions, when other methods of liquidation of goafs are impossible or technologically and economically burdensome. Considering the increase in depth, the use of rockfill has an advantage over hydraulic backfilling, whose installation must be able to withstand high pressure. Very often, to improve the strength parameters of rockfill, binding agents are added [13,14] and waste materials [15–17]. Tesarik et al. [18] and Emad et al. [19] informed about the application of cemented rockfill during pillar extraction in order to increase the stability of excavation.

However, it should be borne in mind that sometimes water from underground workings after the treatment process supplies cities with drinking water [20]. Considering the above criterion, the use of, for example, paste filling with the addition of fly ashes requires additional protection of excavations. In addition, operating costs and the hazard of mine water pollution increase significantly. Problematic issues related to rockfill are associated with, among others: dustiness in the forehead; the need to guarantee a continuous supply of waste rocks with massive use; compressibility, which in some cases can reach close to 50%; and difficulties in organizing transport over long distances.

Wooden crib support is commonly used to strengthen roadway and rooms [21–23]. Zhao [24] presented typical standing supports together with the wooden cribs used for gob-side entry retaining technology. The bearing capacity of a wooden crib increases with the increases of convergence [25,26]. In Polish ore mining, wooden cribs are used to strengthen rooms at the stage of cutting technological pillars and as a local support [27]. In South African underground ore mines depending on the ground conditions, wooden support is widely used in three different support systems: grout packs, crush pillar, and timber props [28]. Erasmus and Smit [29] stated that external support structures, such as timber packs or cribs, form a vital part of the support regime in a reef or seam type of mining operations. Piennar and Howell [30] stated that the new-generation timber packs can safely be used in seismically active areas.

Due to the huge amounts of waste rocks formed in underground ore mining as well as large spaces after exploitation, the author of the article conducted model tests of the three-point and four-point wooden cribs filled with gangue, indicating that they are an excellent way in order to waste management in the mining wooden support.

2. Wood and Cribs in Polish Mining

Wood as a material for supporting underground workings has been known and used in mining practice for a long time. Initially, as the only and basic type of support, it has lost its original meaning over the years, but nowadays it is very often a form of additional protection of workings [31]. Compared to other materials used in underground construction, wood has a number of positive properties, such as compressive strength, bending and flexibility, relatively low volume weight, and easy processing with simple tools, and thus the ability to quickly produce a support structure adapted to local conditions. The wooden support is of particular importance during the reconstruction of the collapses and in all kinds of rescue operations, where there is a need to quickly reach the cut-off areas in the mine through collapsed or endangered with collapses. The advantages of the wooden support include its ability to warn miners about increased pressure in the face, which is manifested by a characteristic crack. Recently, wood has been used in underground construction as an auxiliary material for all kinds of roadways [32], room, and stopes [33], partly in some mining excavations: longwall faces, strip, and room and pillar methods [34]. In shaft construction, wood is used to make inter-level shafts or shallow shafts with a rectangular cross-section. The following types of wood are used in Polish mining: pine, spruce, beech, oak, fir, and larch. The above-mentioned advantages and versatility of use put aside the disadvantages of wood, which include short service life, low durability, susceptibility to decay, and flammability. A common disadvantage of all types of wood is its tendency to absorb moisture and evaporate it when heated, which causes cracks. Increasing the humidity of the wood leads to its rotting, and also has a negative effect on its mechanical strength. Wood is protected against rotting and decaying with a layer of tar, asphalt, or concrete; carbonization of the surface layer by burning; and also by impregnation with special anti-rot preparations [35]. The wood used in the construction of mining support is exposed primarily to compressive and bending forces, and to a lesser extent to shear and tensile forces. The basic strength parameters of wood, i.e., an anisotropic material, are closely related to the direction of external forces. The direction of the force may be parallel, perpendicular to the fiber, radially, and perpendicular to the fiber, tangentially. The compressive strength along the fibers is higher than that across the fibers (Table 1).

Table 1. Basic strength parameters for various types of wood commonly used in Polish underground mines, based on [36].

Type of Wood	Volume Density [kg/m^3]	Compressive Strength [MPa]		Tensile Strength [MPa]		Shear Strength [MPa]	
		Along the Fibers	Across the Fibers	Along the Fibers	Across the Fibers	Along the Fibers	Across the Fibers
Pine	550	43.5	7.5	104	3	10	21
Spruce	470	43	6	90	2.7	6.7	22
Beech	730	53	9	135	7	8	29
Oak	710	47	11	90	4	7.5	27
Fir	450	31	4.5	84	2.3	5.1	27
Larch	690	42	6	107	2.3	9	23

One of the significant problems in the Polish national wood economy is the problem of the increasing shortage of wood every year. Therefore, introducing, among others, in mining technology, the most economical and rational use of this material by extending its service life (by impregnation) or replacing it with other materials (steel, concrete, brick, prefabricated elements, plastics) is fully justified and necessary. The ideal solution to this issue would be to replace wood with appropriate plastic materials, the properties of which would coincide with the advantages of wood and eliminate its negative features. The main obstacle at the present stage of the development of plastics chemistry are economic considerations and the relatively high unit cost of a new material compared to wood [37]. Therefore, wood will be used in mining for many years to come as a valuable and in many cases irreplaceable material.

The history of using wooden cribs in Polish salt mining dates back to the 13th century [38,39]. Full, openwork, empty, or filled cribs fulfill their role for centuries. The support of tourist and sanatorium excavations that occur in the Bochnia and Wieliczka Salt Mines is not only to preserve the stability of the excavation but also to preserve the monument of Polish culture as well as the evidence of the development of Polish mining. Currently, the Wieliczka and Bochnia Salt Mines serve only as a tourist and sanatorium. Wooden support in the form of cribs (Figure 1a–c) and props and long rock bolts made from glass-reinforced plastic constitute the basic method of protection of the salt excavations.

Another example of the wooden support, including wooden cribs, is the zinc and lead Olkusz-Pomorzany mine. Ore deposits in the Olkusz region in Poland occur in the form nest (Figure 2a) and lenses (Figure 2b). For irregularly lying deposits, the exploration drift grid is 100 m. The exploitation is carried out in a layer of ore-bearing dolomites with a thickness of about 40 m, at a depth of 80 to 120 m. The basic ore minerals are galena and sphalerite (Figure 2c). In the Olkusz-Pomorzany mine, room and pillar and also strip methods are used. The orebody is mined using explosives (emulsion materials: emulinit). The length of the blasting holes is from 2 to 3 m. The number of holes on the forehead with a cross-section of 25m^2 is from 25 to 50. The excavations are secured with a resin rock bolt support with a length of 1.6 m. The post-mining areas are filled with a sand hydraulic filling. The average content of zinc and lead in the deposit is 2.6% and 1.3%, respectively. In 2019, production was at the level of 1.6 million tons. In the Olkusz-Pomorzany mine, the thick deposit is divided into horizontal layers parallel to each other with a thickness of 5 to 6 m. In the first layer, the room and pillar is always used, and in the following layers, the strip method. The strip method is used to mine orebodies with a variable thickness and mineralization with the inclination up to 20° and a thickness of up to 6 m. In the exploitation field, operational pillars 35 or 70 m wide are separated, for unidirectional and bidirectional exploitation (Figure 3a). From the raise gallery, the deposit is mined with a strip with a width and height from 5 to 6 m. The strip method is used in two variants with a barrier pillar and a side dam. In the case of roof rocks with a compressive strength above 50 MPa, the barrier pillar variant is used. For this variant, after mining the strip, the barrier pillar is exploited, starting from the end of the strip (Figure 3b). The maximum strip width is 10 m. To eliminate the phenomenon of sand slipping from the next strip, light canvas dams are used. For roof rocks with compressive strength below 50 MPa, the side dam variant is used. After mining a strip, a wooden filling dam is built along the strip sidewall, at intervals of about 0.8 m from the strip sidewall

(Figure 3c). The dam provides additional roof support. In addition to the side dam, a front dam is being built (Figure 3d) with a hole to drain water. Sometimes, wooden cribs are placed next to the sidewall in order to strengthen the room and roadway excavation (Figure 3e). Backfilling begins at the front of the strip and continues continuously until complete filling. Next strips are mined one next to the other, alongside of side wooden dams.

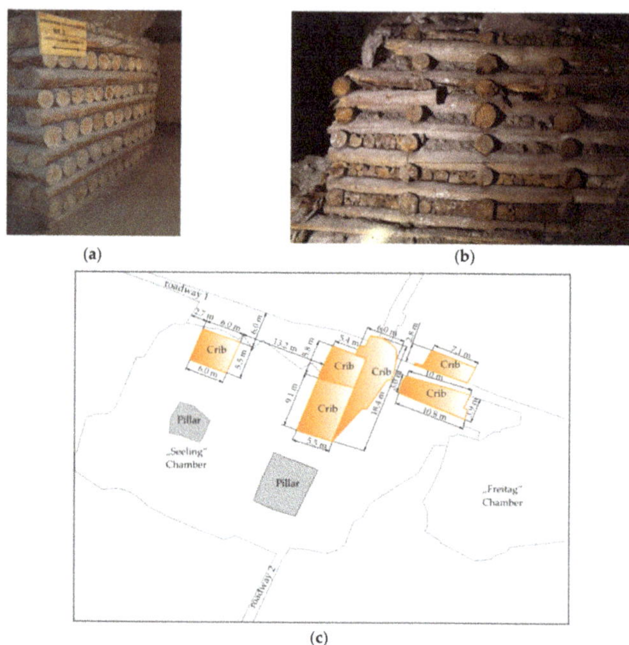

Figure 1. Wooden support of chamber excavations in salt mining: (**a**) Round wooden crib made of round beams, full in the Bochnia Salt Mine; (**b**) Wooden crib filled with gangue in the Wieliczka Salt Mine; (**c**) Protection of "Seeling" Chamber by means of wooden cribs in the Wieliczka Salt Mine.

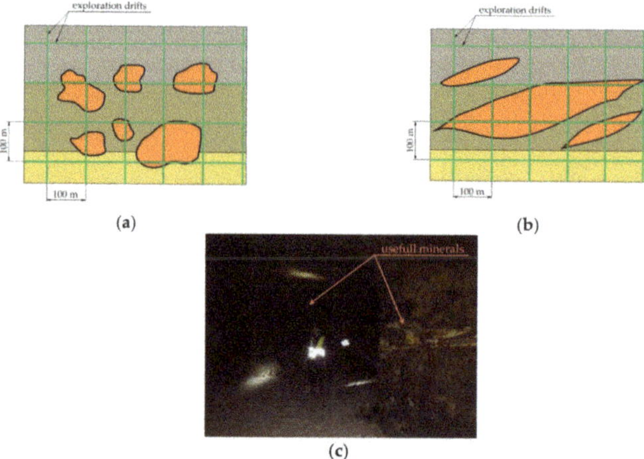

Figure 2. Exemplary forms of deposit in the Olkusz region in Poland: (**a**) Nests; (**b**) Lenses; (**c**) Usefull minerals located in the pillars.

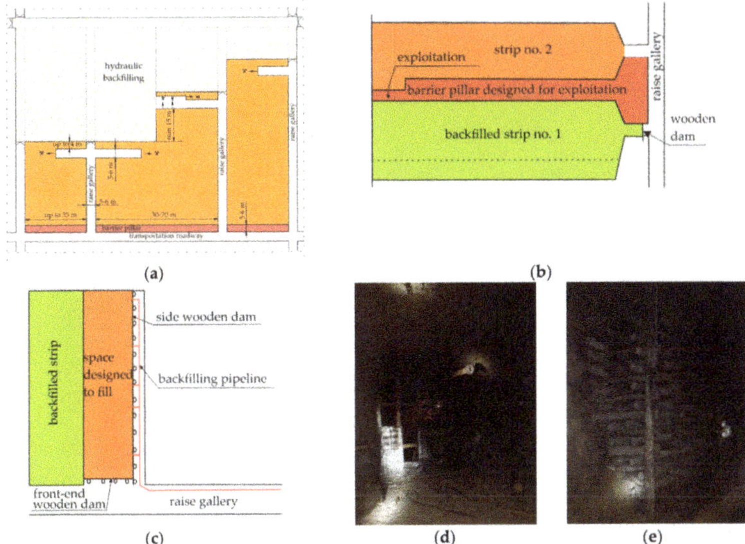

Figure 3. Strip mining method in Olkusz-Pomorzany mine: (**a**) Unidirectional and bidirectional exploitation; (**b**) Exploitation with a barrier pillar; (**c**) Exploitation without a barrier pillar; (**d**) Construction of wooden dam with backfilling pipeline; (**e**) Additional wooden support in a room.

3. Compressibility of the Gangue

In the laboratory tests, gangue was taken from the discharge grate of zinc and lead Olkusz-Pomorzany mine in Poland from a depth of 100 m (Figure 4a). It is worth mentioning that in this mine, rock pieces after blasting have an average diameter of 0.4 m. In order to fill the wooden cribs models, gangue was specially crushed. The average diameter was 0.018 m (Figure 4b). Li et al. [40] and Xiao et al. [41] studied rock material with similar granulation to dry filling. Subsequently, gangue (dolomitic limestone) was placed in oedometers (Figure 4c–d) in order to determine the compressibility. The test results are presented in Figure 5. The internal and outer diameter of the oedometer was 0.113 and 0.150 m, respectively. The diameter and height of the piston was 0.113 and 0.210 m, respectively. The compressibility of the crushed gangues was calculated according to the following formula [42]:

$$C = \frac{h_b - h_a}{h_b} \cdot 100\%,$$ (1)

where:

C—compressibility/%,
h_b—height of the gangues in an oedometer before test/mm,
h_a—height of the gangues in an oedometer after test/mm.

Based on Figure 5, it can be drawn that at a pressure from 0.85 to 1.1 MPa, the compressibility of gangue is 10%. To achieve 20% compressibility, there must be a pressure that ranges from 3.5 to 4.7 MPa. However, the pressure ranges from 11 to 15 MPa corresponds to a compressibility of 30%. At present, exploitation in the zinc and the lead Olkusz-Pomorzany mine is carried out at a depth of 100 m. Assuming that the average unit weight of overburden is 2.7 Mg/m^3, then the vertical pressure is about 2.7 MPa. For this pressure value, the compressibility range is between 15.2% and 18.4%.

Figure 4. Gangue: (**a**) Discharge grate of mining department; (**b**) Jaw crusher with gangue; (**c**) Oedometers filled with gangue; (**d**) Laboratory test of compressibility.

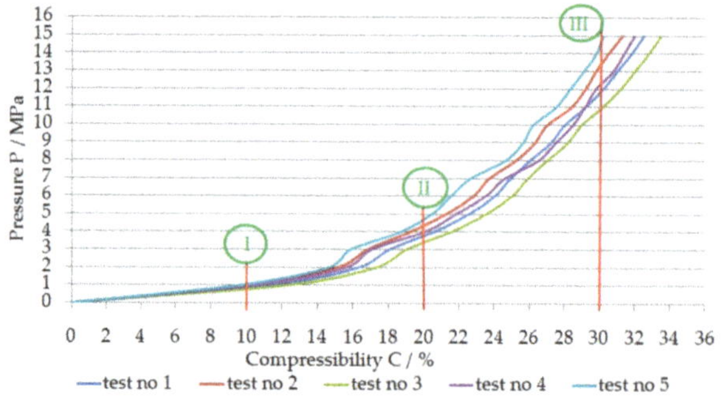

Figure 5. Pressure–compressibility characteristics of gangue (dolomitic limestone): I, II, III—10%, 20%, and 30% of compressibility, respectively.

4. Laboratory Tests of Wooden Cribs Models

Laboratory tests were carried out at the AGH University of Science and Technology, Faculty of Mining and Geoengineering in Poland. The tests were carried out on a hydraulic testing machine with a manual hydraulic oil flow valve. In order to determine the load-displacement characteristics, specialized measuring equipment was used. All sensors were certified and marked by the manufacturer. Three strain gauge force sensors were used to measure the load, while an incremental cable encoder was used to measure the displacement. The individual sensors were connected to a measuring amplifier, which was connected to the computer via an internet cable. The computer was equipped with a specialized CatmanEasy program, with the help of which the load and displacement of the tested crib model were monitored on an ongoing basis. The measurement results were saved in ASCII format and in the Excell program, which was used to draw the load-displacement characteristics. On the laboratory station, three-point and four-point wooden cribs made of pine wooden beams with a triangular and square cross-section were used. The beams of the triangular crib had special notches, made at an angle of 60 degrees. The purpose of the notches was to connect adjacent beams (Figure 6). The individual beams were joined together by quick-drying adhesive. The setting time of the adhesive was 30 s. The height of three-point crib model was 0.180 m while the four-point was 0.150 m. Wooden beams, after appropriate cutting, were dried and had a humidity of 8% to 10%. For the construction of three-point and four-point wooden cribs, models used beams with dimensions of 0.2 m × 0.015 m × 0. 015 m and 0.15 m × 0.015 m × 0.015 m, respectively (Figure 7a–l). Models of three-point and four-point cribs were built of 36 and 20 beams, respectively. In addition, three-point and four-point cribs were filled with gangues (dolomitic limestone) (Figure 7d–f,j–l), with an average diameter 0.018 m. Dolomitic limestone was characterized by compression strength equal to 30 MPa [43]. Only for the four-point wooden cribs models, fiber glass building mesh as an inside lining with a dimension 0.005 m was additionally used.

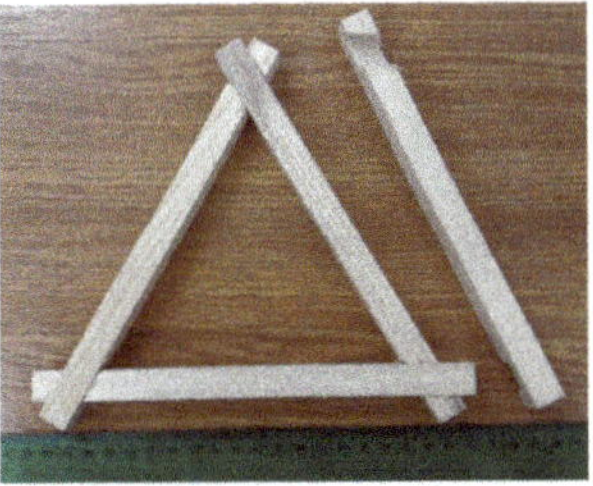

Figure 6. The beams of triangular crib with special notches.

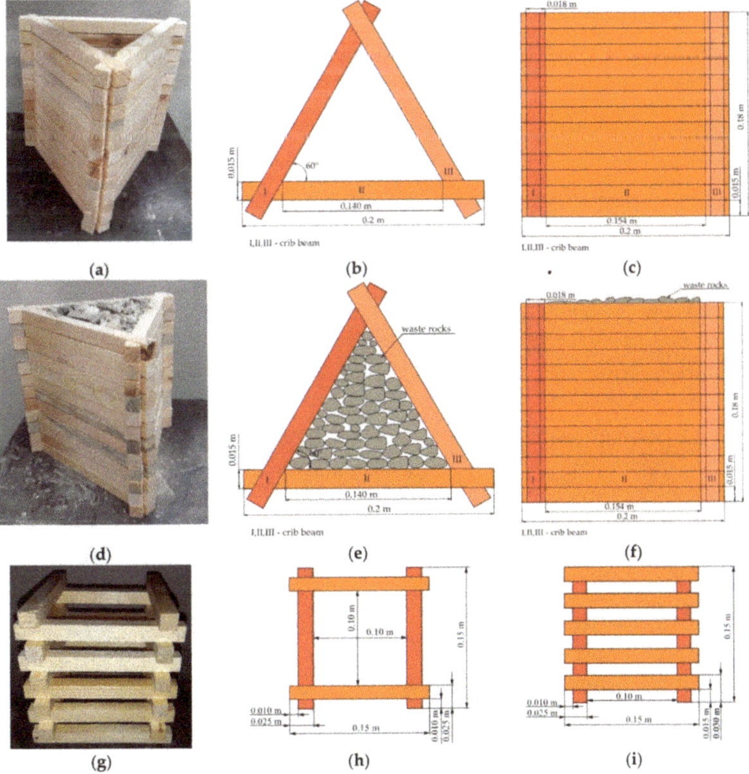

Figure 7. *Cont.*

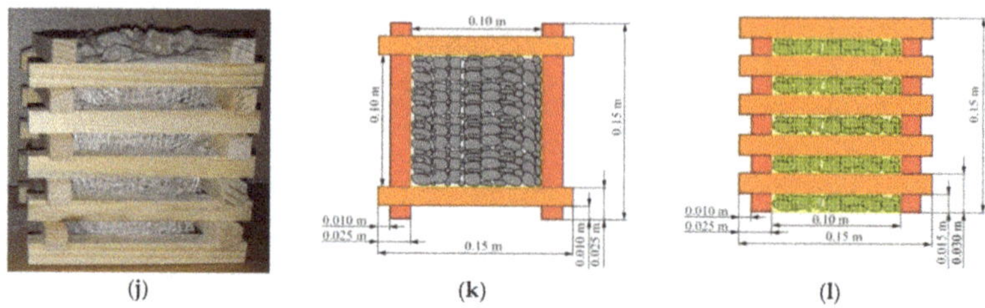

(j)   (k)   (l)

Figure 7. Multi-point cribs: (**a**) Three-point, general view; (**b**) Three-point, top view; (**c**) Three-point, side view; (**d**) Three-point, filled with gangue, general view; (**e**) Three-point, filled with gangue, top view; (**f**) Three-point, filled with gangue, side view; (**g**) Four-point, general view; (**h**) Four-point, top view; (**i**) Four-point, side view; (**j**) Four-point, filled with gangue, general view; (**k**) Four-point filled with gangues, top view; (**l**) Four-point, filled with gangue, side view.

The three-point models did not use an internal mesh as the crib beams were adjacent to each other. The results of the compression tests are presented in Table 2 and in Figure 8a–d and in Figure 9a–d.

Table 2. Summary of the strength tests.

Crib Model	Maximal Load F/kN			Vertical Displacement Δl/mm (at Maximal Load)			Average Specific Strain ε/%
	from	to	Average	from	to	Average	
three-point	33.5	37.9	35.79	40.5	59.9	49.18	27.32
four-point	23	27	25	87.7	99.7	93.26	62.17
three-point filled with gangue	96.7	103.2	100.04	65	80	74.83	41.57
four-point filled with gangue	230	243	235.76	85.7	92.3	89.3	59.53

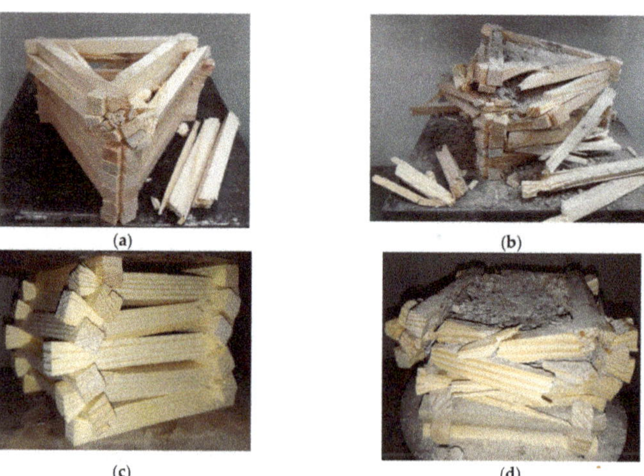

(a)   (b)

(c)   (d)

Figure 8. Wooden cribs models after compression test: (**a**) Three-point; (**b**) Three-point filled with gangue; (**c**) Four-point; (**d**) Four-point filled with gangue.

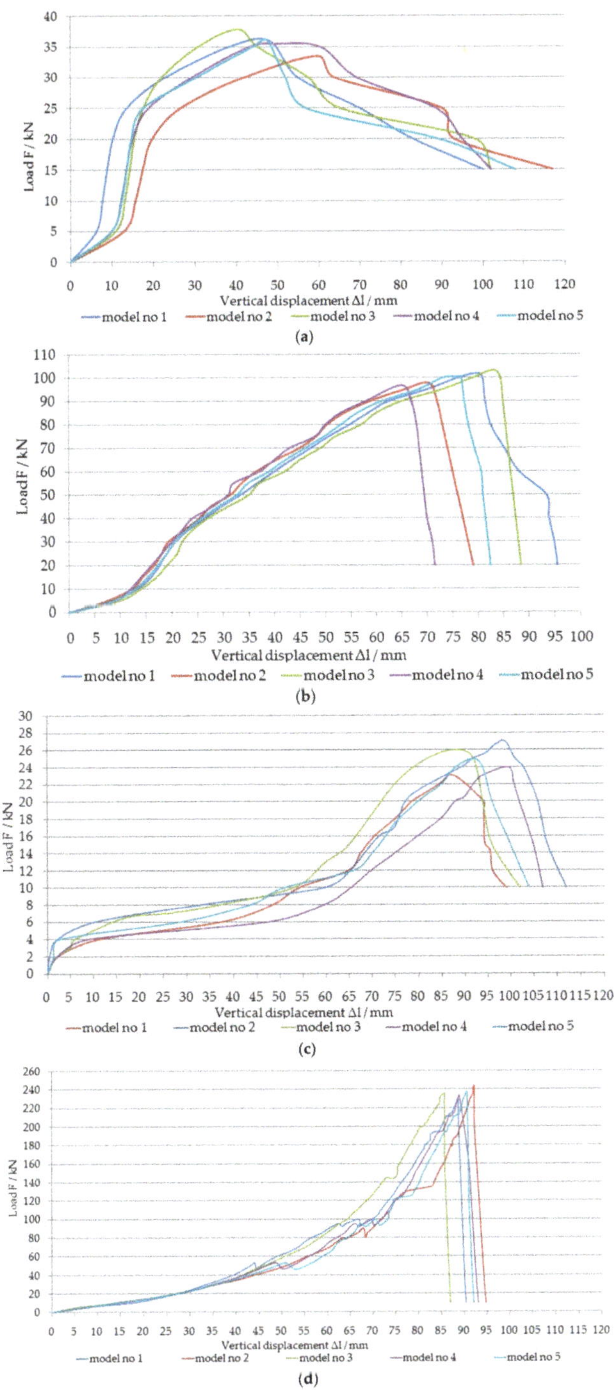

Figure 9. Load-displacement characteristics of crib models: (**a**) Three-point; (**b**) Three-point filled with gangue; (**c**) Four-point; (**d**) Four-point filled with gangue.

Based on Figure 9a–d, it can be drawn that up to maximal load, stiffness for the empty three-point wooden cribs models is 0.72 mm/kN, which means that it is 0.46 higher compared to four-point empty cribs. Comparing empty three-point and four-point cribs with cribs filled with gangue; it can be stated that the stiffness difference is 0.61 and 2.38 kN/mm, respectively. On the other hand, four-point filled models of cribs show much greater stiffness compared to the three-point filled models with gangue; the difference is as much as 1.31 kN/mm. In the process of compression wooden models of cribs, two characteristic failure models can be distinguished; compression until the material is destroyed and compression combined with bending of the crib beam. The first model relates to empty cribs, in which, in the first stage of compression, individual beams are fitted to each other, it is also related to the deletion of clearances between individual beams. Then, the beams of the crib are deformed; the beams press into each other. In the last stage of compression, the crib transfers the load until the wood delaminates and, consequently, leads to breakage of the beams and destruction of the crib structure. The second model of failure relates to the cribs filled with gangue. Compared to the empty cribs, it can be seen that the first compression step consists in eliminating the clearances between the crushed gangue and clamping the crib beams. As a result of the load increase, irregular pieces of the gangue press against the crib beams. In particular, free spaces between the beams are exposed to surface pressure. Due to these pressures, the wood is weakened, the beams break, which in turn leads to the combined process of bending and compressing the crib. This combined process continues until the gangue spills out.

5. Discussion

One the purposes of the research was the use of gangue; therefore, additionally, pressure-compressibility characteristics for three-point and four-point wooden cribs filled with gangue were made in this study (Figure 10a,b). Calculations of pressure and compressibility were made in the Excel spreadsheet by creating a special formula using the results of loading and displacement of the examined cribs. The pressure value was calculated as the ratio of the load to the surface area of the wooden crib models filled with gangue. The surface area was taken into account, in accordance with the dimensions shown in Figure 7e,k. On the other hand, the value of compressibility was calculated as the value of the displacement increase to the initial height of the crib; the result obtained was multiplied by 100%. The value of compressibility was calculated up to the value of 50%, bearing in mind that the three-point cribs achieved compressibility only up to 40%. Above this range, the three-point cribs were completely destroyed. The wooden crib models were tested under static load. The loading rate was 0.5 kN/s.

Based on Figure 10a,b, it can be concluded that for four-point cribs, much less pressure is needed than for three-point cribs. For the specified value of compressibility, the average value of pressure from five measurements was calculated. Then, the obtained results for four-point and three-point cribs were compared with each other, indicating the range of the variability. The difference in favor of the four-point crib is 0.725, 1.99, 2.31 and 1.35 MPa, in relation to 10%, 20%, 30%, and 40% of compressibility, respectively. Based on the obtained results, it can be concluded also that the four-point crib has significantly better strength parameters compared to the three-point crib, for which the fifth compressibility range can be marked (Figure 10b).

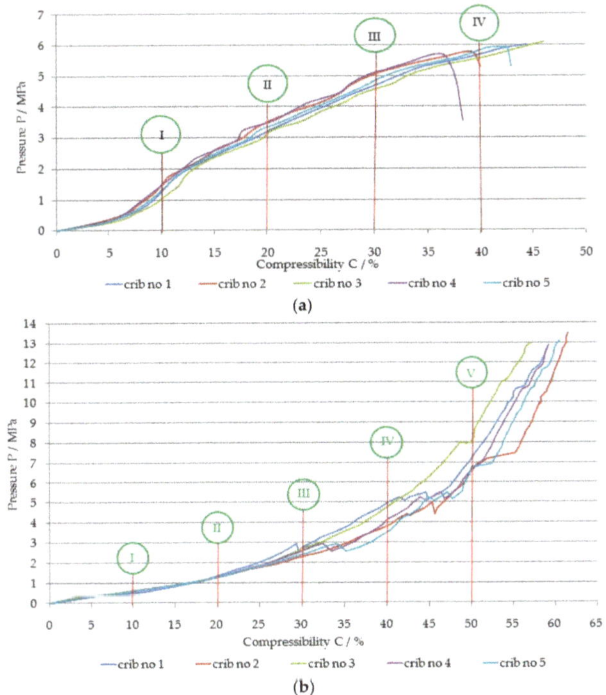

Figure 10. Pressure-compressibility characteristics of wooden crib models filled with gangue: (a) Three-point; (b) Four-point; I–V—compressibility range.

Potential directions for the use of wooden cribs filled with gangue could be used in the strip mining methods in the Olkusz-Pomorzany mine. Due to the shallow exploitation depth of about 100 m, the vertical pressure is about 2.7 MPa. Moreover, taking into account the fact that the exploitation of a part of the deposit is carried out in an undeveloped area and the lack of natural hazards, this means that the post-mining space can be developed with wooden cribs filled with the gangue. On the basis of the obtained results of the compressibility of the wooden cribs, it can be concluded that for the pressure equal to 2.7 MPa, the average compressibility is 17% and 30%, respectively, for the three-point and four-point cribs. Taking into account the fact that in industrial conditions, the value of the cribs' deformation is much greater, and the range of compressibility would also be increased from a few to a dozen percent [25]. This is due to the scale effect of the tested samples [44]. Since the models of four-point cribs filled with gangue were characterized by more than twice the load-bearing capacity (Table 2) compared to three-point cribs, the more favorable model in terms of strength was selected for the analysis of the scale effect. An example of the application of the model tests of wooden cribs in industrial conditions is the underground hard coal mine LW Bogdanka S.A. in Poland. On the basis of the model tests of beech wooden cribs with different filling [45], wooden cribs were built in the mine, which were used to reinforce the maingate behind the longwall face [46]. As a result of the use of additional maingate strengthening, the convergence of the excavation was reduced by 20-30%. In this research, the scale effect was taken into account according to Weibull's theory [47], which shows that the strength of an element depends on its volume, which is expressed by the relationship:

$$\frac{R_1}{R_0} = \left(\frac{V_0}{V_1}\right)^{\frac{1}{m}},$$ (2)

where:

R_1—predicted compressive strength of the crib filled with gangue on a geometric scale 1:1;
R_0—compressive strength of the four-point wooden crib model filled with gangue;
V_0—volume of the four-point wooden crib model filled with gangue;
V_1—predicted volume of the crib filled with gangue on a geometric scale 1:1;
M—Weibull modulus; it can be estimated from the formula [44]:

$$m = \frac{1.2}{C_V},$$

(3)

where C_v is the coefficient of variation of the strength of the material, defined as the ratio of the standard deviation to the average value. It was assumed in this study that the volume of the four-point wooden crib model filled with waste rocks was 0.00150 m^3 (Figure 7k–l); the average compressive strength determined experimentally under laboratory conditions for 10%, 20%, 30%, 40%, and 50% of compressibility is equal to 0.6, 1.4, 2.55, 4.25, and 7.31 MPa, respectively (Figure 10b). Taking into account that vertical stress in the mine is about 2.7 MPa, as the convergence of the excavation increases, the compressibility of the wooden cribs will increase, contributing to the improvement of stability conditions. Currently, in the Olkusz-Pomorzany mine, wooden cribs with round beams, as shown in Figure 3e, are used. Due to the fact that the exploitation in the aforementioned mine is carried out at a shallow depth of about 100 m below the ground surface and that there are no tremors or roof fall rocks in the mine (according to the research carried out by an expert for the support of mining excavations, it is possible to leave the rooms and roadways without a rock bolt support [48]), I performed model tests of wooden cribs indicating the possibility of their future consideration for use. The predicted volume of the wooden crib filled with gangue on a geometric scale of 1: 1 will be 205.78 m^3, a material strength coefficient of variation (for models of cribs filled with gangue and compressive strength of 0.6, 1.4, 2.55, 4.25, and 7.31 MPa, respectively): C_v = 0.196, 0.084, 0.046, 0.083, 0.064, and Weibull modulus m is equal: 6.122, 14.285, 25.97, 14.45, and 18.75, respectively. Taking into account the above data, the predicted compressive strength of the wooden cribs filled with waste rocks on a geometric scale 1: 1 will be 0.09, 0.61, 1.62, 1.87, and 3.88 MPa for compressibility 10%, 20%, 30%, 40%, and 50%, respectively.

Comparing two technologies of liquidation of the post-mining spaces: wooden cribs filled with gangue and hydraulic backfill for the strip method, it can be stated that the total costs for wooden cribs will be significantly lower. The total costs associated with the use of hydraulic backfill can be expressed by the following formula:

$$T_1 = b_1 + b_2 + b_3 + b_4 + b_5 + b_6 + b_6 + b_7 + b_8 + b_9 \; / \; PLN,$$

(4)

where:

b_1—cost of filling material (sand)/PLN;
b_2—cost of transporting of filling material /PLN;
b_3—material cost of the filling pipeline /PLN;
b_4—cost of the medium (water) transporting filling material /PLN;
b_5—cost of construction the filling dams /PLN;
b_6—energy costs associated with pumping out filling water /PLN;
b_7—cost of building wooden dams with canvas /PLN;
b_8—cost associated with taking the gangue to the surface /PLN;
b_9—cost related to the storage of gangue on the surface /PLN.

However, the costs associated with the use of wooden cribs filled with gangue can be expressed by the following formula:

$$T_2 = w_1 + w_2 + w_3 + w_4 \; / \; PLN,$$

(5)

where:

w_1—material cost (wood for building cribs: either fresh or obtained from old railway sleepers)/PLN;
w_2—cost of transporting wooden cribs,/PLN;
w_3—costs of building wooden cribs,/PLN;
w_4—costs of transport of gangue from the mining department to the strips,/PLN.

For wooden cribs filled with gangue, energy costs associated with pumping and draining water will be eliminated. It is worth noting that in the zinc and lead Olkusz-Pomorzany mine, the most energy-consuming and at the same time cost-intensive process is associated with pumping water into surface. The inflow of water is about 200 m^3/min [49]. In order to minimize the costs associated with the use of gangue, it would be necessary to conduct separate mining, that is separate the gangue and orebody with valuable ore minerals, provided that the limitation of the presence of mineralization is clearly marked, as in Figure 2c. In the case of exploratory works in the waste rocks (Figure 2a,b), then there is no need to take it to the surface, but it must be transported directly to the strips.

An exemplary strip mining method for a zinc and lead ore deposit with wooden cribs filled with gangue is shown in Figure 11. The strip mining method can be divided into three stages. In the first stage (Figure 11a), the ore is mined in two directions, leaving a continuous inter-room pillars between the mining rooms. In the second stage, the mined ore deposit is filled with wooden cribs filled with gangue and then the inter-room pillars are mined on both sides (Figure 11b). In the third stage, wooden cribs are again built in the empty spaces and filled with gangue (Figure 11c).

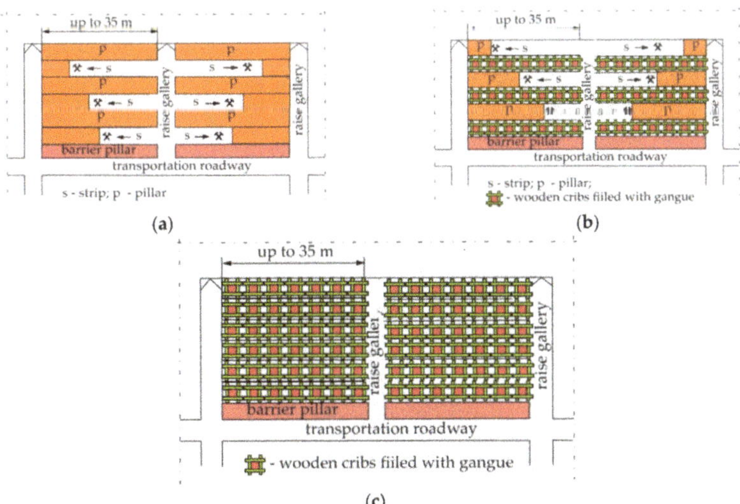

Figure 11. Strip mining method with the three stages of exploitation in relation to the management of wooden cribs filled with gangue: (**a**) First stage; (**b**) Second stage; (**c**) Third stage.

One point to be considered in an inclined orebody is the forces acting on the crib (Figure 12). The values of individual forces were calculated according to the following formulas:

$$F_1 = \gamma \cdot H \cdot (a + 1) \cdot (b + 1') \cdot \cos \alpha, \tag{6}$$

$$F_2 = c \cdot \gamma \cdot H \cdot (a + 1) \cdot (b + 1') \cdot \cos \alpha, \tag{7}$$

$$N_1 = \gamma \cdot H \cdot (a + 1) \cdot (b + 1') \cdot \cos^2 \alpha, \tag{8}$$

$$N_2 = c \cdot \gamma \cdot H \cdot (a + 1) \cdot (b + 1') \cdot \sin 2\alpha, \tag{9}$$

$$T_1 = \frac{1}{2} \cdot \gamma \cdot H \cdot (a + 1) \cdot (b + 1') \cdot \cos^2 \alpha, \tag{10}$$

$$T_2 = \frac{1}{2}\cdot c\cdot\gamma\cdot H\cdot(a+1)\cdot(b+l')\cdot\sin 2\alpha,\tag{11}$$

where:

F_1, F_2—vertical and horizontal force, respectively/N;
N_1, T_1, N_2, T_2—components of vertical and horizontal forces (normal and shear loads), respectively /N;
γ—density of adjacent rocks/N/m^3;
H—foundation depth of the crib/m;
l—strip width along the dip/m;
a—crib width along the dip/m;
l'—strip width along the strike/m;
b—crib width along the strike/m;
α—inclination angle of the orebody/°;
c—horizontal expansion factor (for geological conditions of the Olkusz-Pomorzany mine, c = 0.5).

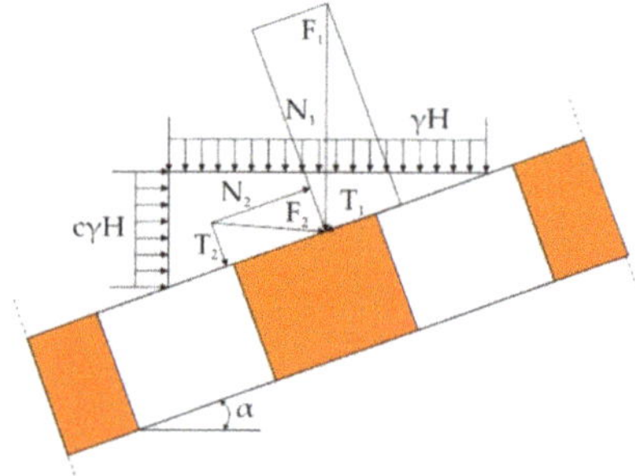

Figure 12. Scheme of loading of cribs in an inclined orebody.

By solving the system of forces and taking into account the surface of the cribs, the formulas for the value of average normal (σ) and shear (τ) stresses are obtained:

$$\sigma = \gamma\cdot H\cdot\frac{S}{s}\cdot\left(\cos^2\alpha + c\cdot\sin^2\alpha\right),\tag{12}$$

$$\tau = \frac{1}{2}\cdot\gamma\cdot H\cdot(1-c)\cdot\frac{S}{s}\cdot\sin 2\alpha,\tag{13}$$

where:

S—roof area per crib /m^2;
S—crib surface /m^2.

The compressive and shear strength of the crib between the strips should meet the condition:

$$\frac{s}{S} \geq \frac{\gamma\cdot H\cdot\left(\cos^2\alpha + c\cdot\sin^2\alpha\right)}{k_c},\tag{14}$$

$$\frac{s}{S} \geq \frac{\gamma\cdot H\cdot(1-c)\cdot\sin 2\alpha}{2\cdot k_t},\tag{15}$$

where:

k_c—allowable compressive stress/MPa;
k_t—allowable shear stress/MPa.

In inclined orebodies, cribs should be tilted from the direction perpendicular to the roof and the floor of the strip towards the rise (Figure 13). Then, the concentration of compressive stresses is reduced and, consequently, the load capacity of the crib increases. The optimal angle of deflection β can be found according to the equation:

$$\beta = \alpha - \text{arctg} \cdot (c \cdot \text{tg}\alpha), \tag{16}$$

where:

β—optimal angle of deflection (for $\alpha = 20°$; $\beta = 10°$)/°;
arctg—inverse tangent;
α—inclination angle of the orebody/°.

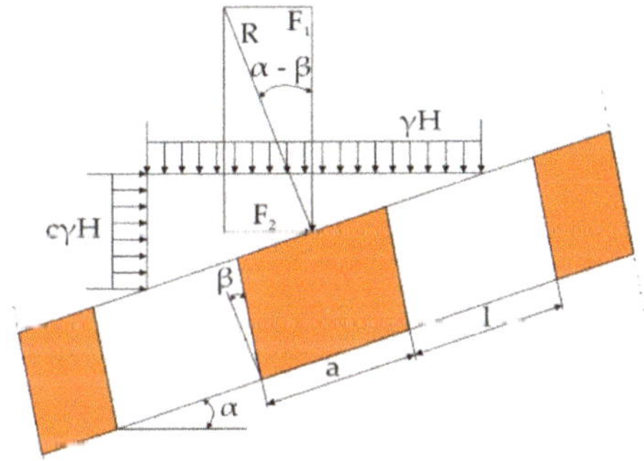

Figure 13. Scheme of loading of cribs in an inclined orebody and inclined from the axis of cribs.

The value of the resultant force (R) can be obtained by solving the system of forces shown in Figure 13:

$$R = c \cdot \gamma \cdot H \cdot S \cdot \frac{\sin \alpha}{\sin(\alpha - \beta)}. \tag{17}$$

Normal stresses in the section perpendicular to the axis of the crib can be calculated according to the formula:

$$\sigma' = \frac{R}{s \cdot \cos \beta} = \gamma \cdot H \cdot \frac{S}{s} \cdot \frac{c \cdot \sin \alpha}{\cos \beta \cdot \sin(\alpha - \beta)}. \tag{18}$$

Assuming that the normal stresses are to be less than the allowable stresses in compression, we get:

$$\frac{s}{S} \geq \frac{\gamma \cdot H \cdot \frac{c \cdot \sin \alpha}{\cos \beta \cdot \sin(\alpha - \beta)}}{k_c}. \tag{19}$$

Analyzing Formulas (12) and (19), it can be seen that they are similar to those for horizontally located orebodies (Formula (20)):

$$\frac{s}{S} \geq \frac{\gamma \cdot H}{C_s}, \tag{20}$$

where:

C_s—compressive strength of the crib /MPa.

Expressions in these formulas:

$$\cos^2 \alpha + c \cdot \sin^2 \alpha = k_\alpha, \tag{21}$$

and:

$$\frac{c \cdot \sin \alpha}{\cos \beta \cdot \sin(\alpha - \beta)} = k_\alpha,\tag{22}$$

where:

k_α—factor of orebody inclination (for square cribs along the strike and along the dip, k_α is equal 0.97 and 0.94 [50], respectively for $\alpha = 20°$).

Inclined cribs show less load capacity than horizontal cribs. The dimensions of the inclined cribs can therefore be determined by the formulas given for horizontal orebodies, introducing into them the factor of orebody inclination k_α, determined by Formulas (21) and (22).

6. Conclusions

Based on the laboratory research, it can be drawn:

- Three-point empty wooden crib models have a higher load capacity of over 30% compared to the four-point crib.
- Filling the three-point crib made of 36 beams filled with gangue increases its load capacity by more than twice.
- Filling the four-point crib made of 20 beams filled with gangue increases its load capacity by over nine times.
- The four-point wooden crib filled with gangue has a load capacity greater by 57% compared to the filled three-point wooden crib.

Underground ore mining continues to progress to deeper levels. The transport of excavated output and waste rocks contributes to increased production costs. A solution that has been used for a long time in mining technologies is the management of waste rock in both exploitation and preparatory excavations. At the intersection of excavations, where sidewalls are damaged very often, using three-point or four-point wooden cribs can be a quick and effective method of ensuring the stability of the excavation. Comparing the two types of crib connections, it should be stated that in mining conditions, the best solution would be to use four-point cribs due to the simple way of laying the beams. Moreover, considering the possibility of monitoring the load on the cribs, the four-point crib also has an advantage over the three-point crib as it is possible to place the sensor between the lower and upper crib beam at any height. Additionally, by filling the crib with waste rock, its condition can be macroscopically observed. Taking into account the amount of gangue that could be utilized in the crib (Figure 7e,k), it should be stated that the four-point crib creates much more favorable conditions. Its surface area for gangue management is 58% larger compared to the three-point crib. This means that to use the same amount of gangue, almost twice as many three-point cribs as four-point cribs should be used. The obtained results of laboratory tests may be helpful in making a decision on the method of securing the excavation with the simultaneous placement of gangue.

Funding: This research was prepared as part of AGH University of Science and Technology in Poland, scientific subsidy under number: 16.16.100.215.

Conflicts of Interest: The author declares no conflict of interest.

References

1. Chang, Q.; Chen, J.; Zhou, H.; Bai, J. Implementation of Paste Backfill Mining Technology in Chinese Coal Mines. *Sci. World J.* **2014**, *2014*, 1–8. [CrossRef] [PubMed]
2. Zhang, J.; Li, M.; Taheri, A.; Zhang, W.; Wu, Z.; Song, W. Properties and Application of Backfill Materials in Coal Mines in China. *Minerals* **2019**, *9*, 53. [CrossRef]
3. Yildiz, T.D. Waste management costs (WMC) of mining companies in Turkey: Can waste recovery help meeting these costs? *Resour. Policy.* **2020**, *68*, 1–17. [CrossRef]

4.	Yang, G.-L.; Yang, R.-S.; Tong, Q.; Huo, C. Coalmine Green Mining with Gangue Backfilling Technique. *Procedia Environ. Sci.* **2011**, *10*, 1205–1209. [CrossRef]

5.	Li, H.; Guo, G. Surface Subsidence Control Mechanism and Effect Evaluation of Gangue-Backfilling Mining: A Case Study in China. *Geofluids* **2018**, *2018*, 1–9. [CrossRef]

6.	Sun, G.; Feng, T.; Liu, J. Survey on roof movement of the gangue backfill working face. *J. Vibroeng.* **2019**, *21*, 1107–1115. [CrossRef]

7.	Cao, W.; Wang, X.; Li, P.; Zhang, D.; Sun, C.; Qin, D. Wide Strip Backfill Mining for Surface Subsidence Control and Its Application in Critical Mining Conditions of a Coal Mine. *Sustainability* **2018**, *10*, 700. [CrossRef]

8.	Tai, Y.; Han, X.; Huang, P.; An, B. The mining pressure in mixed workface using a gangue backfilling and caving method. *J. Geophys. Eng.* **2019**, *16*, 1–15. [CrossRef]

9.	Ran, J.J. Safe mining practices under wide spans in underground non-caving mines—Case studies. *Int. J. Min. Sci. Technol.* **2019**, *29*, 535–540. [CrossRef]

10.	Yin, Y.; Zhao, T.; Zhang, Y.; Tan, Y.; Qiu, Y.; Taheri, A.; Jing, Y. An Innovative Method for Placement of Gangue Backfilling Material in Steep Underground Coal Mines. *Minerals* **2019**, *9*, 107. [CrossRef]

11.	Skrzypkowski, K. Case Studies of Rock Bolt Support Loads and Rock Mass Monitoring for the Room and Pillar Method in the Legnica-Głogów Copper District in Poland. *Energies* **2020**, *13*, 2998. [CrossRef]

12.	Skrzypkowski, K.; Korzeniowski, W.; Zagórski, K.; Zagórska, A. Modified Rock Bolt Support for Mining Method with Controlled Roof Bending. *Energies* **2020**, *13*, 1868. [CrossRef]

13.	Lingga, B.A.; Apel, D.B. Shear properties of cemented rockfills. *J. Rock Mech. Geotech. Eng.* **2018**, *10*, 635–644. [CrossRef]

14.	Jiang, H.; Fall, M.; Li, Y.; Han, J. An experimental study on compressive behaviour of cemented rockfill. *Constr. Build. Mater.* **2019**, *213*, 10–19. [CrossRef]

15.	Skrzypkowski, K.; Korzeniowski, W.; Poborska-Młynarska, K. Binding capability of ashes and dusts from municipal solid waste incineration with salt brine and geotechnical parameters of the cemented samples. *Arch. Min. Sci.* **2018**, *63*, 903–918. [CrossRef]

16.	Korzeniowski, W.; Poborska-Młynarska, K. Skrzypkowski, K. The idea of the recovery of municipal solid waste incineration (MSWI) residues in Kłodawa Salt Mine S.A. by filling the excavations with self-solidifying mixtures. *Arch. Min. Sci.* **2018**, *63*, 553–565. [CrossRef]

17.	Liu, W.; Xu, J.; Zhu, W.; Wang, S. A novel short-wall caving zone backfilling technique for controlling mining subsidence. *Energy Sci. Eng.* **2019**, *7*, 2124–2137. [CrossRef]

18.	Tesarik, D.R.; Seymour, J.B.; Yanske, T.R. Long-term stability of a backfilled room-and-pillar test section at the Buick Mine, Missouri, USA. *Int. J. Rock Mech. Min.* **2009**, *48*, 1182–1196. [CrossRef]

19.	Emad, M.Z.; Mitri, H.; Kelly, C. Dynamic model validation using blast vibration monitoring in mine backfill. *Int. J. Rock Mech. Min.* **2018**, *107*, 48–54. [CrossRef]

20.	ZGH Bolesław. Available online: https://www.zghboleslaw.pl (accessed on 21 September 2020).

21.	Timrite. Available online: http://www.timrite.co.za//products (accessed on 29 July 2020).

22.	Stratawordwide. Available online: https://www.strataworldwide.com/timber-roof-supports (accessed on 29 July 2020).

23.	Miningproducts. Available online: http://www.miningproducts.co.za/mining-products.html (accessed on 29 July 2020).

24.	Zhao, H. State-of-the-art of standing supports for gob-side entry retaining technology in China. *J. S. Afr. Inst. Min. Metall.* **2019**, *119*, 891–906. [CrossRef]

25.	Skrzypkowski, K. Decreasing mining losses for the room and pillar method by replacing the inter-room pillars by the construction of wooden cribs filled with waste rocks. *Energies* **2020**, *13*, 3564. [CrossRef]

26.	Korzeniowski, W.; Skrzypkowski, K.; Szumiński, A.; Zagórski, K. Laboratory investigations of the load capacity and load-strain characteristics of four-point cribs filled with waste rocks. *Bull. Inst. Non-Ferrous Metals.* **2015**, *60*, 303–308. [CrossRef]

27.	Skrzypkowski, K. The influence of dimension of mining crib model filled by waste rock on the course of load–displacement characteristic. *Bull. Min. Energy Econ. Res. Inst. Pol. Acad. Sci.* **2017**, *99*, 131–141.

28.	Ozbay, M.U.; Ryder, J.A.; Jager, A.J. The design of pillar systems as practiced in shallow hard-rock tabular mines in South Africa. *J. S. Afr. Inst. Min. Metall* **1995**, *1/2*, 7–18.

29. Erasmus, P.N.; Smit, J. Assessment of Precast Cellular Lightweight Concrete (C.L.C.) Support Structures. In Proceedings of the International Symposium on Ground Support; Villaescusa, E., Windsor, C.R., Thompson, A.G., Eds.; Taylor & Francis Group: Kalgoorlie, Australia; Rottedram, The Netherlands, 1999; pp. 349–357.

30. Pienaar, F.R.P.; Howell, M. *The Rapid Loading of Packs. SANIRE Free State Symposium 2007. Redefining the Boundaries Part II*; South African National Institute of Rock Engineering: Johannesburg, South Africa, 2007; pp. 1–10.

31. Skrzypkowski, K. Strengthening the room excavation through the use of additional support in the form of wooden props in underground ore copper mines. *Ores Non-Ferr. Met. Recycl.* **2017**, *5*, 10–15.

32. Brown, C.J. Development process for a higher capacity Propsetter® system. *Int. J. Min. Sci. Technol.* **2018**, *28*, 121–126. [CrossRef]

33. Malan, D.F.; Napier, J.A.L. Rockburst support in shallow-dipping tabular stopes at great depth. *Int. J. Rock Mech. Min.* **2018**, *112*, 302–312. [CrossRef]

34. Klemetti, T.M.; Sears, M.M.; Tulu, I.B. Design concerns of room and pillar retreat panels. *Int. J. Min. Sci. Technol.* **2017**, *27*, 29–35. [CrossRef]

35. Strzelecki, Z. *A Material Science Guide for Underground Construction of Mines*; Śląsk Publishing House: Bytom, Poland, 1972; p. 21.

36. Bańkowski, Z. *Mechanic's Guide*; First Part; Scientific and Technical Publishers: Warszawa, Poland, 1988; p. 567.

37. Solis, M.; Silveira, S. Technologies for chemical recycling of household plastics—A technical review and TRL assessment. *Waste Manag.* **2020**, *105*, 128–138. [CrossRef]

38. The Bochnia Salt Mine. Available online: https://bochnia-mine.eu (accessed on 4 September 2020).

39. The Wieliczka Salt Mine. Available online: https://www.wieliczka-saltmine.com (accessed on 4 September 2020).

40. Li, M.; Zhang, J.; Sun, K.; Zhang, S. Influence of Lateral Loading on Compaction Characteristics of Crushed Waste Rock Used for Backfilling. *Minerals* **2018**, *8*, 552. [CrossRef]

41. Xiao, Y.; Meng, M.; Daouadji, A.; Chen, Q.; Wu, Z.; Jiang, X. Effects of particle size on crushing and deformation behaviors of rockfill materials. *Geosci. Front.* **2020**, *11*, 375–388. [CrossRef]

42. Skrzypkowski, K. Compressibility of materials and backfilling mixtures with addition of solid wastes from flue-gas treatment and fly ashes. *E3S Web Conf.* **2018**, *71*, 1–6. [CrossRef]

43. Skrzypkowski, K. The influence of room and pillar method geometry on the deposit utilization rate and rock bolt load. *Energies* **2019**, *12*, 1–15. [CrossRef]

44. Zweben, C. Designer's corner. Is there a size effect in composites? *Composites* **1994**, *25*, 451–454. [CrossRef]

45. Korzeniowski, W.; Skrzypkowski, K. Comparative investigations of the load capacity and load-strain characteristics of chocks with different filling. *Mining Review.* **2012**, *68*, 36–40.

46. Korzeniowski, W.; Herezy, Ł.; Krazue, K.; Rak, Z.; Skrzypkowski, K. *Rock Mass Monitoring Based on Analysis of Powered Support Response*; AGH Publisher: Kraków, Poland, 2013; pp. 153–165. ISBN 978-83-7464-554-6.

47. Weibull, W. A statistical distribution function of wide applicability. *J. Appl. Mech.* **1951**, *18*, 293–297.

48. Regulation of the Minister of Energy of November 23, 2016 on detailed requirements for the operation of underground mining plants. *Journal of Laws of the Republic of Poland*, 9 June 2017; Item 1118.

49. *Hydrogeological Documentation of the Olkusz-Pomorzany Mine*; Mining and Metallurgy Plant: Bukowno, Poland, 2020.

50. Piechota, S. *Technique of Underground Mining of Deposits*; AGH Publisher: Kraków, Poland, 1988; p. 196.

Article

A Multi-Criteria Approach for the Evaluation of Low Risk Restoration Projects in Continuous Surface Lignite Mines [†]

Philip-Mark Spanidis [1], Christos Roumpos [2],* and Francis Pavloudakis [2]

[1] ASPROFOS Engineering, Division of Project Management, 17675 Athens, Greece; pspani@asprofos.gr
[2] Mining Engineering Department, Public Power Corporation of Greece, 10432 Athens, Greece; f.pavloudakis@dei.com.gr
* Correspondence: c.roumpos@dei.com.gr; Tel.: +30-6979799291
† This paper is an extended version of our paper published in "Spanidis, Philip-Mark P. & Roumpos, Christos & Pavloudakis, Francis. (2018). A Multi-Criteria Methodology for Low-Risk Evaluation of Mine Closure Restoration in Continuous Surface Lignite Mining Projects, in Conference: 14th International Symposium of Continuous Surface Mining, At Thessaloniki, Greece".

Received: 21 March 2020; Accepted: 28 April 2020; Published: 1 May 2020

Abstract: The restoration of continuous surface lignite mines entering the closure phase is a long-term, complex and multidisciplinary project. During the evaluation of alternative restoration technologies, various environmental, technical, economic and social parameters are investigated. In this framework, for the selection of the lower risk restoration alternative, the analysis of the associated risks should be incorporated into the decision-making process. This work provides an overview of practical risk management problems and solutions in mining restoration projects. Moreover, it introduces a multi-criteria methodology for the improvement of the decision-making process in the evaluation of restoration alternatives and the selection of the optimal one, considering a continuous surface mining project. The proposed method is a combination of the analytical hierarchy process (AHP) for the quantification of risk factors and the technique for order of preference by similarity to ideal solution (TOPSIS) for the ranking of restoration alternatives based on a low risk approach. The results of the case study indicate that the proposed approach can be utilized as a low cost and easy-to-apply tool, appropriate for coal mining operators, managers and stakeholders involved in the planning and implementation of post-mining land restoration activities. Furthermore, the suggested methodology could be adopted to support the risk management needs in the restoration stage of complex surface mining projects.

Keywords: decision making; management; mines; multi-criteria analysis; project; risk; restoration

---

1. Introduction

Surface lignite mines exploited using a continuous mining method are typically large-scale excavation sites of high significance for primary energy production. The lignite extracted from these mines fuels high capacity thermal power plants. Although the operational life of a lignite mine usually spans many decades, after an intensive exploitation period, the content of lignite basins is depleted. Then, the lignite mines enter the closure phase that necessitates restoration of the post-mining sites, with beneficial effects to environment and society.

Continuous surface mining projects are very complicated because of the uncertainties associated with their dynamic situation. In this type of projects, the risks may relate to geological, technical, environmental, social, economic and other factors. In this framework, strategic mine planning, aimed at the rational exploitation of the coal/lignite deposits, is crucial for the sustainability of such projects. This planning should incorporate the restoration of mining sites, considering the optimal mine development and the related risks as well as the time and cost data of mining activities. It requires, therefore, an integrated approach and optimization based on technical, economic, environmental, safety or social parameters.

Post-mining land restoration is related to integrated and long term practices of high social, economic and environmental impact. Literature reports various restoration methods. Some are more technical, dealing with engineering solutions for mitigating environmental impacts of the post-mining sites [1,2]. Others focus on restorations mainly achieved by natural processes, known as 'spontaneous succession' [3] or suggest hybrid or 'near natural' solutions [3,4]. Most of the methods cited consider reclamation as a fundamental activity related to extended earthworks for the removal of the excavated waste material (mainly consisting of interburden, overburden and topsoil material) and subsequent use of as much volume of this material as possible for the filling of excavation pits, voids and other open trenches shaped during the mine life cycle [2,5,6]. Based on the rules of aesthetics as well as of landscape architecture, the reclaimed areas should be perfectly integrated towards landscape reformulation and upgrading [7].

The planning of mining restoration is a multidisciplinary task requiring the synergies of managers, engineers, environmentalists, risk analysts and experts from various fields of science and technology, as well as local stakeholders (i.e., independent parties/bodies with interests or concerns in the approval and licensing of restoration projects). The main planning objective is to select the techno-economically feasible and low risk sustainable restoration methods that are the most advantageous for each site-specific case in terms of economic growth, prosperity of local society and ecosystem balance. In this context, the spatial analysis of selected indicators and cartographic overlay using ArcGIS applications can provide a useful tool, either for evaluating the corresponding sustainable exploitation of a mineral deposit [8] or for selecting the optimum land uses that should be developed in the reclaimed lands after the mine closure [9–11].

The surface mine restoration projects are associated with numerous risks and uncertainties related to natural and technological parameters [12,13]. As a result, the decision making for the selection of the most advantageous alternative is a difficult and complicated process. Therefore, the early identification of restoration risks and their impacts enables decision makers to have a substantial basis of understanding the advantages and disadvantages of each alternative. In turn, they will be able to select the optimal one which combines the lower risks with the higher expected benefits for the environment, society and economy. This is still a challenging research topic in the field of optimization, restoration and sustainability of surface coal mines.

The development of MCDM methods and techniques is a growing trend worldwide. Recently, a considerable literature has grown up around this field, covering a wide spectrum of academic research and practical applications in the industry, such as project and risk management, business and finance, manufacturing, etc. [14–17] From this spectrum, the mining industry should not be an exception [18]. Taking into account that the mining projects are long term and large scale frameworks with high inherent complexity, the needs of solving problems regarding the production optimization, land management and reclamation planning are critical and require substantial, cost effective, environmentally friendly and regulatory compliance decisions.

In the mining industry, various MCDM techniques and associated research have been developed in the recent decades. According to Mahase et al. [18], those techniques are applied to support, in principle, decision making needs for mine planning and related problems. Sousa Jr. et al. [19] used the PROMETHEE and ELECTREE methods in a process for the selection of highway trucks for mining operations. Musingwini [20] and Musingwini and Minnitt [21] ranked the efficiency of certain mining methods using the analytical hierarchy process (AHP) technique. Relevant to the mining methods selection are also the works of Bitarafan and Ataei [22] and De Almedida et al. [23] Yavouz [24] proposed a method for equipment selection using a combination of Yager's method and AHP. For the selection of the most appropriate underground mining method, Karadogan et al. [25]

presented the application of fuzzy set theory in a case for the selection of an underground mining method. Kazakidis et al. [26] suggested the application of the AHP in a series of different case studies/scenarios to investigate the investment analysis of a new mining technology, the design aspects for the ground support, the design of tunneling systems, the shaft location selection and the mine planning risk assessment. Khakestar et al. [27] performed a multi-criteria evaluation based on a hybrid combination of the fuzzy AHP and the technique for order of preference by similarity to ideal solution (TOPSIS) techniques for the selection of the optimum method regarding the slope stability analysis in open pit lignite mines. Bazzazi et al. [28] assessed a combination of the AHP and TOPSIS techniques to select optimal loading-haulage equipment for the Sungun open pit mine in Iran. Bascetin [29] recommended a decision support system using the AHP technique for the optimal reclamation of an open-pit mine. Finally, Uberman and Ostrega [30] applied the AHP in a project selection for the revitalization of post-mining areas.

The results of the above investigation indicate several interesting findings:

(a) The MCDM methods are mainly focused on needs for improvement of the mining operations and equipment supply, while the post-mining restoration activities seem to be of a lower priority.
(b) A preference for the development of decision making methods as combinations (or "hybrids") of two, or more, MCDM techniques is noted.
(c) The AHP tends to be the fundamental constituent element in most of the hybrid techniques and, especially, for the decision making step, where the weights of the criteria are determined.
(d) References regarding the assessment of risk-based decision making for the restoration of continuous surface mines are limited.

In this context, the aim of this paper is to propose a new approach to the research field of planning the land reclamation of surface mines by (a) demonstrating the decision-making problems and risks related to the selection of an appropriate post-mining restoration technology/method and (b) suggesting a low risk methodology based on multi-criteria decision making (MCDM) techniques, aiming to better understand and control the risks in surface mining restoration projects. Within this framework, the paper introduces a hybrid MCDM methodology [31–35], which is a combination of AHP (analytical hierarchy process) [36–38] and TOPSIS (technique for order of preference by similarity to ideal solution) [34,35], and fills a gap in the literature regarding the quantitative risk assessment in the field of mine land reclamation. A case study approach was used for the implementation of the methodology considering the selection of the lower risk restoration/reclamation method in a continuous surface lignite mining project.

The paper is organized as follows: the second section provides background information on restoration planning and the related risks and impacts. The third section presents practical decision-making problems and questions. The fourth section describes the suggested MCDM methodology. The fifth section demonstrates the implementation of the methodology in a case study. The final section provides conclusions, and directions for future research.

2. Background

In many countries, such as Germany [1], the Czech Republic [2], India [5], UK [39], the US [40] and Greece [41], the restoration of mining sites and industrial areas is a regulatory requirement. In this frame, the restoration of continuous surface lignite mines requires a multidisciplinary approach and properly planned activities of returning the mined-out lands to a form of acceptable environmental condition and productivity [5], along with beneficial returns to society and the economy. Nevertheless, the analysis of mining industry practices reveals that there are several restoration approaches, since the restoration does not follow a standard conceptual model, because of the significant differences in mining conditions. The exploitation models depend on the specific physical and operational characteristics of each mine. However, the post exploitation situation depends, mainly, on the geo-environmental features of the mined-out sites, as well as on the environmental legislation and regulatory constraints. Empirical evidence and literature deals with various restoration approaches, however, three representative technologies can be, in principle, figured out: (a) technical restoration [2,42], (b) natural restoration driven by interactions between various bio-ecological and non-biotic factors [1] and (c) combined restoration consisting of partial human intervention developed in parallel to the natural restoration processes [2,4,6].

The basis for a techno-economically effective and environmentally friendly restoration plan is the successful reclamation of mined-out lands [1,2,5,6]. Reclamation is a large scale reinstatement activity aimed at returning the mined-out site to its original form, productivity and land use. Thus, reclamation enables landscape remodeling and, being conducted with other human activities and/or natural restoration processes, it contributes to the recovery of areas affected by the intensive operations of mine exploitation. The reclamation works during mining operations are mainly related to the extended backfilling of open pits, benches, open trenches and voids shaped by the mining activities.

From the project management viewpoint, restoration may be seen as a sub-project in a wider environmental management program, which maintains its own budget, execution schedule, organization, human resources, equipment and materials. Once a restoration project is under planning, managers, engineers and restoration experts are working in synergy to outline, in an appropriate level of integration, a multidisciplinary project definition report. For this purpose, they decompose and analyze thoroughly the geo-environmental conditions of the mining landform characteristics, as well as the techno-economic profile of the restoration project. In this analysis, critical parameters for the identification of restoration methods are assessed, i.e., quality of disposed soils, lakes and rivers, air pollution, topsoil fertility, infrastructure (roads, railways and welfare facilities), abandoned mining machinery, toxicity of discarded wastes, reforestation, recreation, geo-hazards, land use parameters, etc. The data required for the restoration analysis are collected through field investigations across the mined-out sites, water and soil laboratory analyses, satellite imagery and cartographic/GIS analysis products [2,43], as well as through review of statistical records, scientific literature and technical documents of the mine facility. The synthesis of a comprehensive restoration plan, recommending several technically adequate restoration methods and suggesting the most advantageous one for implementation, is a common practice in the evaluation phase of such projects. The plan is also delivered to the involved parties and stakeholders (environmental agencies, investors, municipalities, NGOs, etc.) for consideration and decision making.

On the other hand, the long term mine restoration projects, with time frames varying from few years to decades, are related to various risks. These risks have to be identified as early as possible, and an assessment of the impacts should be conducted. This work is critical for the determination of the overall risk associated with each alternative restoration method, in order to evaluate and finally select the lower risk alternative as appropriate and technically adequate for implementation.

A thorough investigation of the fundamental restoration risks has been reported in the work of Roumpos et al. [12], however, a further analysis on this topic exceeds the scope of this paper. The main risks are grouped in five categories, while each category is further divided into sub-categories as depicted in Tables 1–5: technical risks, geological and geotechnical risks, permitting risks, socio-economic risks and environmental risks.

Table 1. Surface Lignite Mines Restoration Technical Risks (R1).

Notation	Description
R1.1	Landscape formation incompatible with existing infrastructures (roads, railways, buildings, etc.)
R1.2	Re-handling of disposed waste material.
R1.3	Restrictions due to remaining/exploitable mineral resources
R1.4	Mine closure requirements before the exhaustion of the reserves
R1.5	Failures of excavation machinery
R1.6	Inappropriate mine decommissioning plan (for large facilities, plants and excavation equipment)

Table 2. Surface Lignite Mines Restoration Technical Geological and Geotechnical Risks (R2).

Notation	Description
R2.1	Landslides, slope instabilities and soil deformations
R2.2	Pit lakes stability problems
R2.3	Subsidence phenomena
R2.4	Hydrogeological management problems
R2.5	Inappropriate reinstatement of the terrain relief and topography
R2.6	Seismicity and tectonics

Table 3. Surface Lignite Mines Restoration Permitting Risks (R3).

Notation	Description
R3.1	Delays of the permission process
R3.2	Deviations of the reclamation plan from development plans approved by authorities
R3.3	Conditions, technologies and activities of high cost and complexity requested by authorities
R3.4	Constraints due to claims and requests addressed by the stakeholders

Table 4. Surface Lignite Mines Restoration Socio-Economic Risks (R4).

Notation	Description
R4.1	High capital expenditures (CAPEX) of reclamation works (project)
R4.2	High maintenance cost of reclaimed lands
R4.3	High cost of environmental monitoring and control measures
R4.4	Requirements for land acquisition
R4.5	Utilization of low quality and productivity soil(s)
R4.6	Protests of local communities against the reclamation plans
R4.7	Low returns to local and regional economy/society
R4.8	Low priority of reclamation plan, according to a strategic mine closure plan
R4.9	Visual amenity and aesthetic impacts
R4.10	Lack of technical and economic capabilities for beneficial modification of the post mining area
R4.11	Low value of reclaimed land

Table 5. Surface Lignite Mines Restoration Environmental Risks (R5).

Notation	Description
R5.1	Requirements for re-exploitation of already restored mines area
R5.2	Effects on groundwater, surface water, noise and air quality
R5.3	Loss and/or alteration of biodiversity and biotopes
R5.4	Post-mining natural hazards and area disturbances
R5.5	Post-mining drainage problems
R5.6	Effects to flora, fauna and vegetation
R5.7	Effects to surrounding habitats/ecosystems
R5.8	Legal constraints for the management of mining by-products (ash, sludge)
R5.9	Incompatibility with the surrounding environment and landscape
R5.10	Low soil fertility
R5.11	Remaining pit voids
R5.12	Unavailability of topsoil material
R5.13	Misalignment of the restoration/reclamation plan from the objectives of sustainable development

3. Decision Making: Problems and Questions

The restoration of continuous surface lignite mines are projects of high technical, environmental and socioeconomic complexity, since the post-mining sites cover hundreds of square kilometers. In addition, restoration activities usually last for many years. The success of the restoration projects depends on the quality of restoration planning and decisions taken in the pre-investment phase. In fact, the improper evaluation of restoration plans might lead to unsuccessful decisions to accept inappropriate plans as feasible or, conversely, to reject restoration plans which in a deeper level of analysis are feasible and sustainable. Therefore, the decision making for the evaluation and selection of the most advantageous restoration technology for implementation is very critical, as it is directly related to whether a restoration project can be successfully implemented, delayed or may even lead to failure. Therefore, the quality of decisions is crucial and constitutes a vital element that indicates whether a restoration project can be implemented or not, particularly in the pre-investment phase, where the project feasibility and planning is subject to investigation.

The mining practice shows that the significant differences observed between restoration projects introduce various technical and managerial problems. Consequently, several considerations and screenings for the primary geo-environmental factors and reclamation technologies or methods advised by the restoration experts may be conflicting. For instance, a simple and low cost reclamation technology in sites polluted by toxic discarded materials might generate new environmental impacts of high severity. Moreover, the inappropriate restoration activity may relate to land use incompatibilities in a way that the restored land could become inappropriate for reforestation, recreation or re-cultivation. This fact might be subject to rejection by environmentalists, permitting agencies, socioeconomic analysts or stakeholders.

There are various decision making practices with respect to project management. Each practice depends on the nature of the project, the collective performance of experts and managers, environmental sustainability policies and regulations, budgetary and resource constraints and other geo-environmental factors. Some practices are very formal and move within the standards of administrative and legal protocols of public agencies and organizations involved with the regulatory compliance of mining restoration plans or wider programs. Other more empirical and subjective practices are based on mining experts' judgement and rely on comparisons with previous projects of similar scale, type and techno-economic profile. Some practices examine combinations of empirical evidence of the geo-environmental and ecological conditions of the under restoration areas, while others focus on the perspectives of restoration returns by considering the restoration project as a pure business case. In this view, project managers and restoration experts, in order to achieve an integrated and substantial approach to select the most adequate alternative technology/method of restoration, are called to solve various technical problems and to answer critical questions, such as:

- Which restoration technology/method is environmentally friendly, sustainable and cost-effective against others?
- Which technology/method should be appropriate for obtaining the approval of the authorities?
- Which technology/method requires the minimum modifications of existing infrastructures?
- Which technology/method ensures efficient soil improvement, landscape remodeling and visual amenity?
- Are the environmental, ecological and biodiversity requirements completely met?
- Do the alternative technologies/methods satisfy needs for new businesses opportunities, reduction of unemployment and enhancing the livelihood of local communities?
- Are the alternatives appropriate for the recovery of biodiversity, reforestation, replantation and recultivation?
- Are the water management proposals suitable for the development of recreational facilities?
- Are the measures for protection against natural hazards properly investigated?
- Are the methods of toxic wastes removal and water treatment ensuring people's safety/health?

- Are the required mechanical means and resources available for the execution of the restoration earthworks?
- Are there any possibilities for restart, if necessary, of the mining activities in the future?

As the list of relevant questions is long, it becomes obvious that there are many open points which, by their nature, generate significant and multilateral risks. Since the main decision making objective is the selection of the restoration alternative with the lower risk, it is suggested that these risks should not be simply considered as a contingency element in the project investment analysis. Instead, it is suggested that the decision makers should perform a substantial analysis of the restoration risks, where the risks are grouped and classified into specific categories. In this analysis, the risks can be quantified in a form of numerical data and used as criteria for the evaluation of the restoration alternatives. The quantification of risks can be obtained through properly structured questionnaires, where evaluators/experts can express their own perception on the significance of each risk, based on their professional experience. To this regard, the alternative technologies/methods can be evaluated, in terms of an MCDM methodological approach, considering the relative risks. Thus, the 'total risk score' for each alternative can be determined in a numerical form and the alternative with the lower total risk score can be selected as the most appropriate.

4. Suggested Methodology

The suggested methodology focuses on the application of MCDM methods and techniques [44], highlighting their utility as a tool for supporting decision making, regarding the evaluation of post mining restoration projects. The main objectives of the methodology are the following:

(a) To interpret the restoration risks in terms of evaluation criteria and to show how the relative weight of every individual criterion/sub-criterion can be expressed in numerical form.

(b) To show how any alternative restoration technology/method can be decomposed in a risk-based classification matrix in accordance with a pattern for the classification of risk impacts as low, medium or high severity.

(c) To present how the relative weights of the identified criteria/sub-criteria can be used in combination with the risk and sub risk factors to calculate the overall risk of each alternative and, thus, to rank the alternatives.

(d) To prove the applicability of the methodology in a case study regarding the restoration project of a continuous surface mining operation in Northern Greece that enters the closing phase.

The methodology combines two widely known MCDM techniques: the AHP, applied for the quantitative expression of risk-based criteria and sub-criteria, and TOPSIS, applied for the calculation of the overall risk of each alternative and the ranking of the alternatives according to their total risk scores. In the AHP technique, experts use their knowledge, insight and professional experience to analyze the decision-making problem, by defining hierarchy levels and to solve it. In this framework, they follow an evaluation process based on quantified criteria. The selection criteria and sub-criteria in the restoration project of the case study are the relative weights of the identified risk and sub-risk factors. The decision makers perform pairwise comparisons, taking into account the 1–9 rating scale advised by Saaty [36–38,45], to evaluate the criteria and to structure and normalize a reciprocal pairwise comparison matrix. The computational process is verified by a mathematical consistency check and the relative weights of risk factors and sub-factors are finally identified. TOPSIS is based on the assumption that the most advantageous alternative should have the shortest distance from the ideal solution and the highest distance from the negative ideal solution (this distance is also known as 'Euclidean entity'). In this context, the distances of each alternative from the ideal and the negative alternative, respectively, enable the final ranking and classification of all alternatives [32,33,35]. Figure 1 describes the suggested methodology as a process model, based on the combination of the AHP and TOPSIS techniques.

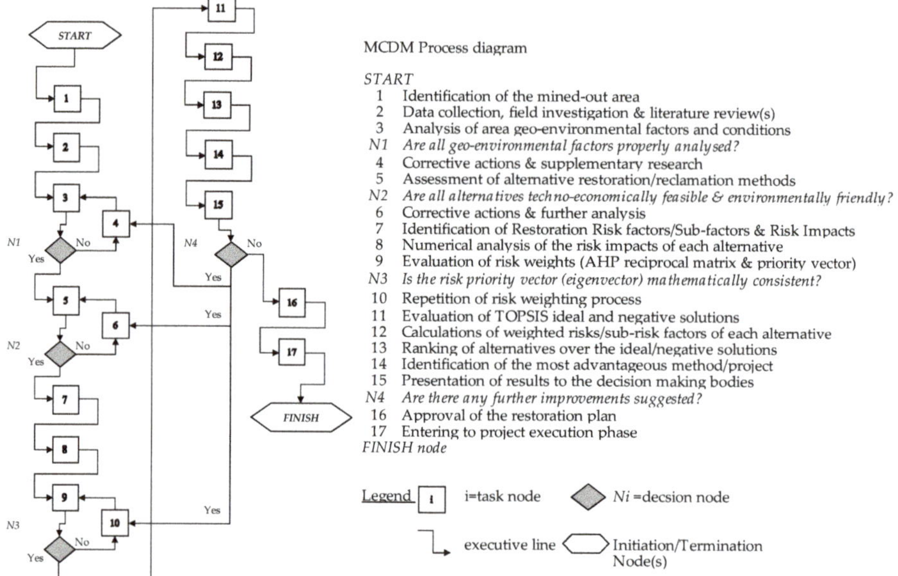

Figure 1. Suggested Methodology.

5. Implementation

5.1. Case Study: Restoration Project Overview

The case study refers to a restoration project of a continuous surface mine, including two lignite fields (Mine I and Mine II, Figure 2) planned for the exploitation of a lignite deposit located in northern Greece. Between the two mining fields a ridge is found with no lignite occurrence. The mining site is hilly, with surface altitudes ranging between 700 m–900 m above sea level. At the western boundary of the mining area, there is a village with 100 inhabitants. The mine contributes in meeting the lignite demand of a power plant located 12 km west of the mining area.

The lignite basin has a channeled form and it is broadened from the NE to the SW direction, along with local increases of surface elevations. The area of mining activity covers approximately 10 km^2. The deposit is of a multiple-layered form, and lignite seams are almost horizontally bedded. However, a series of normal faults results in a progressive and systematic deepening of the beds and in a corresponding increase the thickness of overburden material towards the SE rim of the basin. The main characteristic of the deposit is the occurrence of conglomerate hard material in the overburden formations.

The strategic mine development, including the two outside dumping areas, is shown in Figure 3. The main outside waste dumping area is located 1.5 km NW from the mine. The sequence of mining operations, considering time periods of 10 years, from the initial to the mine closure stage, is presented in Figure 4.

The main mining equipment consists of bucket wheel excavators, conveyors and spreaders and the decision making regards the optimal restoration of the mine after 25 years of operation, while its remaining life is 10 years. At this stage, the waste material of the excavation activities is transported to the inside dumping area. Following the mine closure plan, a lake was planned to fill the existing area voids.

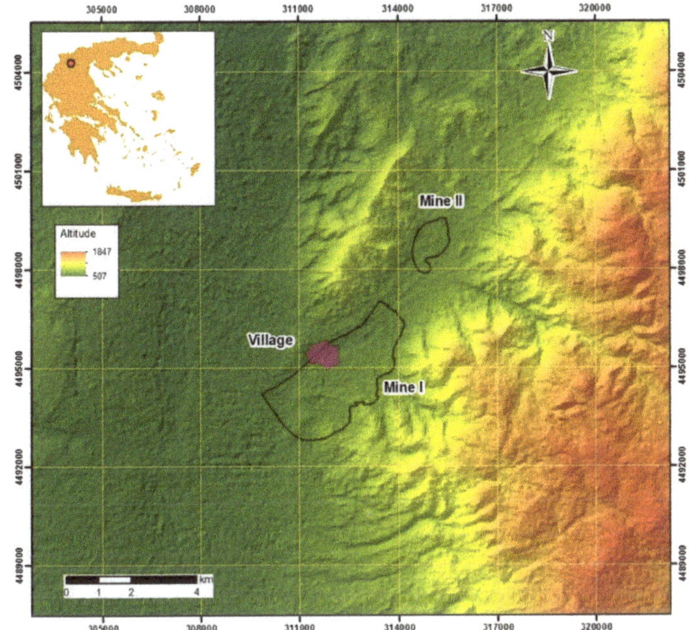

Figure 2. General overview of the mining area, including the boundaries of the two lignite fields (Mine I and Mine II).

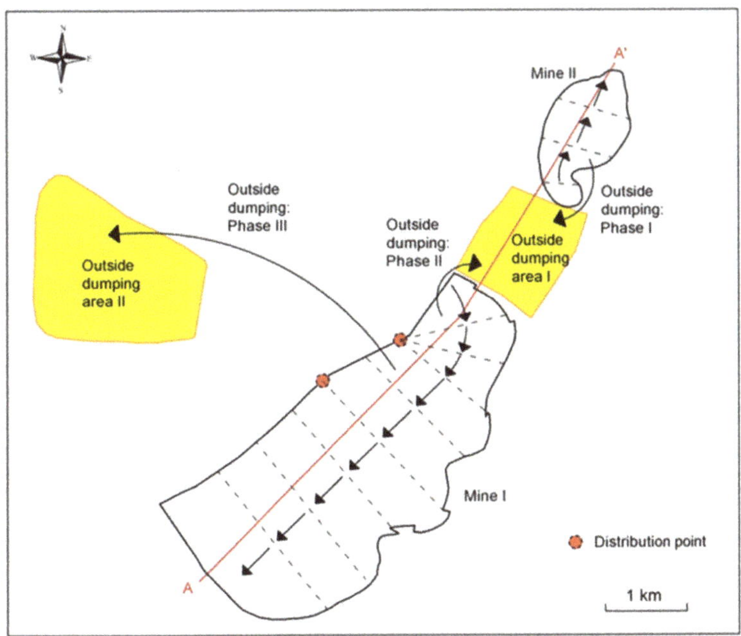

Figure 3. Strategic mine development.

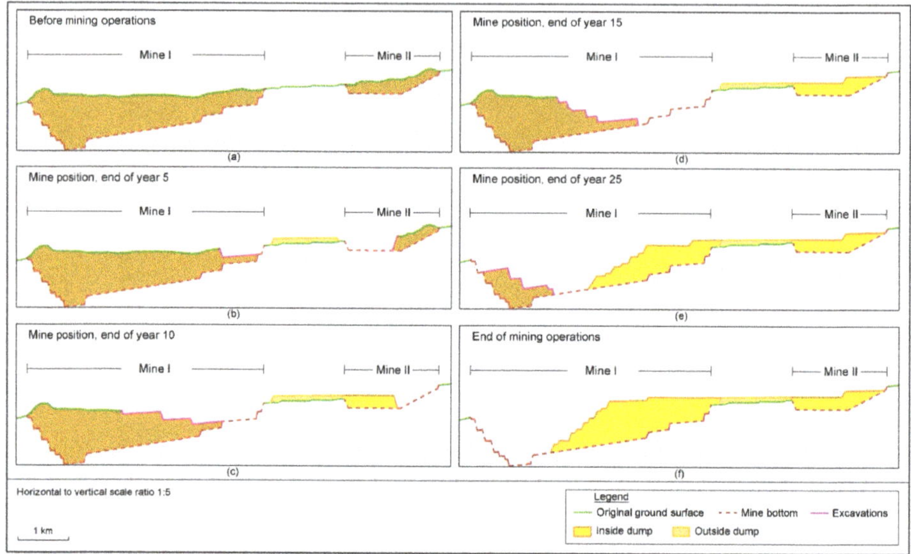

Figure 4. Sequence of mine development in the cross section A-A′ of Figure 2.

5.2. Description of Alternatives

The following three (3) restoration alternatives were examined (the comparative cost data were collected from [46–49]):

(A1) Technical Restoration: It refers to a project with high capital expenditures (3.500–5.000 €/ha) and an implementation period of 3–4 years. The project targets are: (a) large-scale reclamations for recovering the affected landscape, (b) treatment of contaminated, polluted or toxic soils and waters, (c) measures and infrastructures enabling development of recreation activities, (d) upgrading of biodiversity at post mining ecosystems. The measures and main infrastructures suggested are: slope stabilization; excavation/backfilling of removed earth material; land compaction/relevelling; redesign and reconstruction of existing roads; erection of buildings for accommodation of 300 visitors; erection of recreation and cultural facilities (sport camps and a museum for the mining history); reforestation/replantation to the 35–40% of the entire project area; upgrading of near lake settlement; installation of fire protection system; utilities for irrigation, sewing, water supply, power and communication; removal of polluted, contaminated and toxic wastes and charged waters; replacement of selected topsoil; measures ensuring land fertility; measures for recovery of sensitive habitats and biota; measures for increase of employment.

(A2) Restoration focused on Natural Processes: It refers to a project with low capital expenditures (less than 500 €/ha) and an implementation period of 10 years. The project targets are: (a) medium to low scale reclamations for recovering the affected landscape, (b) treatment of contaminated, polluted or toxic soils and waters, (c) protection of biodiversity at post mining ecosystems. The measures and main infrastructures suggested are: selective backfilling at the most adversely excavated lignite layers and seams; soil and stabilization at locations with the highest possibility of post mining erosion effects; removal of polluted, contaminated and toxic soil, mining wastes and water bodies; installation of fire protection system; irrigation and water management measures; measures for recovery of sensitive habitats and biota; replacement of selected topsoil at locations with recultivation capability and good fertility; implementation of a long-term environmental monitoring and physical restoration plan (also known as 'spontaneous succession' [5]) plan.

(A3) Combined Solution: It refers to a project with moderate capital expenditures (800–1500 €/ha), with an implementation period of 5–7 years. The project targets are: (a) large-scale reclamations

for recovering the affected landscape, (b) treatment of contaminated, polluted and toxic soils and waters, (c) measures and infrastructures enabling mainly the development of agroforestal activities, (d) upgrading of the biodiversity at post mining ecosystems. The measures and main infrastructures suggested are: excavation and backfilling of removed earth material; soil and slope stabilization; compaction and levelling; local modifications of existing traffic network; erection of low scale recreation and cultural facilities (sport camps and a museum of mining history); reforestation/replantation of the 25–35% of the project area; installation of fire protection system; removal of polluted, contaminated and toxic soil, mining wastes and water bodies; replacement of removed/selected topsoil at the 40–50% of the project area; measures for recovery of sensitive habitats and biota; implementation of a long-term environmental monitoring and physical restoration plan.

5.3. Risk-Based Analysis

The risk severity impacts are presented in Table 6, reflecting the classification of risks and the units/parameters used to measure the risk severity. In turn, the risk severity is classified into three main categories: low, medium and high, along with the numerical values. Based on these values, the severity of any impact can be determined (the risk impact measuring units were identified by empirical evidence and by considering the specific geo-environmental conditions of the mine restoration plan that were described in the Section 5.1). These values are based on the PMI [50] practice, not to follow a linear order for the escalation of risk severity, but to apply the relation $a(n) = 2^n$ ($n \in N_0$). Thus: $2^0 = 1$: low impact, $2^1 = 2$: medium impact and $2^2 = 4$: high impact [51] (Table 7).

Table 6. Classification of Risk Severity Impacts.

Risk-ID	Units & Parameters of Risk Evaluation	Low (L)	Medium (M)	High (H)	Impact L-M-H
R1	**Technical Risks**				
R1.1	Length of interfered infrastructure [km^2]	none	one (1)	more than one (1)	1-2-4
R1.2	Disposed waste in million [m^3]	less than 1 mil [m^3]	1–2 mil [m^3]	more than 2 mil	1-2-4
R1.3	[%] of the proven mine reserves [m^3]	less than 10%	10–20%	more than 20%	1-2-4
R1.4	[%] of the proven mine reserves [m^3]	less than 10%	10–20%	more than 20%	1-2-4
R1.5	Safety incidents [per month]	none	one (1)	more than one (1)	1-2-4
R1.6	[%] delay of total time-plan [months]	less than 10%	10–20%	more than 20%	1-2-4
R2	**Geological and Geotechnical Risks**				
R2.1	Cases of instability/[km^2]	up to 2 per [km^2]	2–5/[km^2]	more than 5/[km^2]	1-2-4
R2.2	Cases of subsidence/[km^2]	up to 2 per [km^2]	2–5/[km^2]	more than 5/[km^2]	1-2-4
R2.3	[%] of the reclamation area with slopes >25%	none	slopes < 10%	slopes >10%	1-2-4
R2.4	Area [%] of watercourses/[km^2]	up to 2%	2–5%	more than 5%	1-2-4
R2.5	[%] of the reclamation area [km^2]	up to 5%	5–10%	more than 10%	1-2-4
R2.6	Seismic zone (*)	Zone-I	Zone-II	Zone-III	1-2-4
R3	**Permitting Risks**				
R3.1	Approval time [months]	less than 6 months	6–12 months	more than 12 months	1-2-4
R3.2	Number of discrepancies [%]	less than 5	5–10	more than 10%	1-2-4
R3.3	[%] increase of reclamation budget [€]	less than 5%	5–10%	more than 10%	1-2-4
R3.4	Impact of Stakeholders claims	less than 5%	5–10%	more than 10%	1-2-4
R4	**Socio-Economic Risks**				
R4.1	CAPEX of the reclamation project [€/ha]	less than 1000 €/ha	1000-3000 €/ha	more than 3000 €/ha	1-2-4
R4.2	[%] of reclamation CAPEX [€]	less than 5%	5–10%	more than 10%	1-2-4
R4.3	[%] of reclamation CAPEX [€]	less than 5%	5–10%	more than 10%	1-2-4
R4.4	[%] of reclamation area [km^2]	less than 2%	2–5%	more than 5%	1-2-4
R4.5	[%] of reclamation area [km^2]	less than 2%	2–5%	more than 5%	1-2-4
R4.6	Number of discrepancies	less than 5	5–10	more than 10	1-2-4
R4.7	[%] of reclamation CAPEX [€]	less than 5%	5–10%	more than 10%	1-2-4
R4.8	Transition period [months]	less than 4 months	4–10 months	more than 10 months	1-2-4
R4.9	[%] of the reclamation area [km^2]	less than 5%	5–10%	more than 10%	1-2-4
R4.10	Missing resources in [%] of CAPEX [€]	less than 5%	5–10%	more than 10%	1-2-4
R4.11	[%] of the land value of unaffected areas in the greater mining area [€]	less than 60%	60–90%	more than 90%	1-2-4

Table 6. *Cont.*

Risk-ID	Units & Parameters of Risk Evaluation	Low (L)	Medium (M)	High (H)	Impact L-M-H
R5	*Environmental Risks*				
R5.1	Probability of reexploitation	less than 0.15	0.15–0.30	more than 0.30	1-2-4
R5.2	Increase of pollution potential [%]	less than 1%	1–5%	more than 5%	1-2-4
R5.3	Deterioration of biodiversity indices in relation with unaffected areas	more than 30%	10–30%	less than 10%	1-2-4
R5.4	Probability of post-mining hazards	less than 0.05	0.05–0.10	more than 0.10	1-2-4
R5.5	[%] of the reclamation area [km^2]	less than 5%	5–10%	more than 10%	1-2-4
R5.6	[%] of the reclamation area [km^2]	less than 10%	10–20%	more than 20%	1-2-4
R5.7	[%] of the reclamation area [km^2]	less than 10%	10 - 20%	more than 20%	1-2-4
R5.8	[%] of mining by-product volume	less than 10%	10 - 20%	more than 20%	1-2-4
R5.9	[%] of the reclamation area [km^2]	less than 20%	20 -40%	more than 40%	1-2-4
R5.10	[%] of reclaimed area	more than 70%	70- 35%	less than 35%	1-2-4
R5.11	[%] of the reclamation area [km^2]	less than 35%	70- 35%	more than 70%	1-2-4
R5.12	[%] of the required topsoil volume [m^3]	more than 70%	70- 35%	less than 35%	1-2-4
R5.13	Criticality of non-sustainable deviations	low criticality	medium criticality	high criticality	1-2-4

(*) According to the new anti-seismic regulation of Greece

Table 7. Risk-Based Analysis.

Risk-ID	Units & Parameters of Risk Evaluation	Alternative A1 - Rv(1, j)	Alternative A2 - Rv(2, j)	Alternative A3 - Rv(3, j)
R1	*Technical Risks*			
R1.1	Length of interfered infrastructure [km^2]	4	1	2
R1.2	1 million [m^3]	4	1	2
R1.3	[%] of the proven mine reserves [m^3]	2	1	2
R1.4	[%] of the proven mine reserves [m^3]	2	2	2
R1.5	Safety incidents [per month]	4	1	2
R1.6	[%] delay of total time plan [in months]	4	2	2
R2	*Geological and Geotechnical Risks*			
R2.1	Cases of instability/[km^2]	2	4	2
R2.2	Cases of subsidence/[km^2]	2	4	2
R2.3	[%] of the reclamation area with slopes	2	4	2
R2.4	Area [%] of watercourses/[km^2]	4	2	2
R2.5	[%] of the reclamation area [km^2]	1	4	2
R2.6	Seismic Zone (*)	2	1	1
R3	*Permitting Risks*			
R3.1	Approval time in months	4	2	1
R3.2	Number of discrepancies [%]	4	2	1
R3.3	[%] increase of reclamation budget [€]	4	2	1
R3.4	Impact of stakeholders claims	2	4	2
R4	*Socio-Economic Risks*			
R4.1	CAPEX of the reclamation project [€/ha]	4	1	1
R4.2	[%] of reclamation CAPEX [€]	4	2	1
R4.3	[%] of reclamation CAPEX [€]	1	4	2
R4.4	[%] of the reclamation area [km^2]	4	1	2
R4.5	[%] of the reclamation area [km^2]	1	4	2
R4.6	Number of discrepancies	2	4	2
R4.7	[%] of reclamation CAPEX [€]	1	4	2
R4.8	Transition period [months]	4	1	2
R4.9	[%] of the reclamation area [km^2]	1	4	2
R4.10	Missing resources in [%] of CAPEX [€]	2	1	2
R4.11	[%] of the land value of unaffected areas in the greater mining area [€]	2	1	1

Table 7. *Cont.*

Risk-ID	Units & Parameters of Risk Evaluation	Alternative A1 - Rv(1, j)	Alternative A2 - Rv(2, j)	Alternative A3 - Rv(3, j)
R5	*Environmental Risks*			
R5.1	Probability of reexploitation	2	2	2
R5.2	Increase of pollution potential [%]	1	4	2
R5.3	Deterioration of biodiversity indices in relation with unaffected areas	1	2	1
R5.4	Probability of post-mining hazards	1	2	1
R5.5	[%] of the reclamation area [km^2]	1	4	2
R5.6	[%] of the reclamation area [km^2]	2	4	2
R5.7	[%] of the reclamation area [km^2]	2	4	2
R5.8	[%] of mining by-product volume	1	4	1
R5.9	[%] of the reclamation area [km^2]	2	4	1
R5.10	[%] of reclaimed area	2	4	2
R5.11	[%] of the reclamation area [km^2]	1	2	1
R5.12	[%] of the required topsoil volume [m^3]	2	2	2
R5.13	Criticality of non-sustainable deviations	1	4	1
			106	67

(*) According to the new anti-seismic regulation of Greece

5.4. Risk Factors Definition

The definition of the relative weight of each criterion has been performed by applying the AHP technique according to the following steps [32,37,38,45]:

- *Step 1*: Establishment of a pairwise comparison matrix for weighting the main risks; the experts compose a square matrix consisted of two triangular sub-matrices by making pairwise comparisons of each criterion with each one of the other criteria using the Saaty scale of comparison (Table 8). Calculation mode: if a(i, j) is an element of this matrix (k: column and l: row) the lower diagonal element is produced using the formula a(k, l)*a(l, k) = 1 (Table 9);
- *Step 2*: Normalization of the pairwise comparison matrix and extracting the relative weight of each criterion, WRi | i = 1, 2,.., n, n ∈ N; n is the number of criteria; WR(i) represents the priority, or eigenvalue, vector (Table 10); mathematical conditions: $0 < WRi < 1$ and $\sum_{i=1}^{n} WRi = 1$;
- *Step 3*: Performing the consistency control to validate the consistency of priority vector; the control aims to check if the consistency ratio CR < 0.10; if so, the data of priority vector is appropriate for further utilization as inputs in TOPSIS calculations. Otherwise, the risk weighting process is repeated;
- *Step 4*: Steps similar to the above 1–3 and consistency controls are carried out to produce the priority vectors WR(i, j) of each group of sub-criteria (sub-risks).

Table 8. Scale of Criteria comparison [29].

1: Equal importance
3: Moderate importance of one criterion over another
5: Strong or essential importance
7: Very strong importance
9: Extreme importance
2, 4, 6, 8: Values for inverse comparison

Table 9. Analytical hierarchy process (AHP) Reciprocal Matrix.

Main Risks		R1	R2	R3	R4	R5
Technical	*R1*	1	2/3	2/3	3/5	1/2
Geological/Geotechnical	*R2*	3/2	1	1/2	1/3	1/2
Permitting	*R3*	3/2	2	1	1/2	1/3
Socio-Economic	*R4*	5/3	3	2	1	3
Environmental	*R5*	2	2	3	1/3	1

Table 10. Normalized Matrix and Priority Vector, *WR$_i$* (Risk Weights).

Main Risks		R1	R2	R3	R4	R5	WRi
Technical	*WR$_1$*	0.13	0.08	0.09	0.22	0.09	*0.110*
Geological/Geotechnical	*WR$_2$*	0.20	0.12	0.07	0.12	0.09	*0.120*
Permitting	*WR$_3$*	0.20	0.23	0.14	0.18	0.06	*0.161*
Socio-Economic	*WR$_4$*	0.22	0.35	0.28	0.36	0.56	*0.349*
Environmental	*WR$_5$*	0.26	0.23	0.42	0.12	0.19	*0.260*
-		1.00	1.00	1.00	1.00	1.00	1.000
-		Consistency Control:					
	CI =	0.086	RI =	1.12	CR =	0.041	< 0.10

5.5. Ranking of Alternatives

The application of TOPSIS aims to define the score of each alternative and the final ranking of the alternatives and, hence, to demonstrate the lower risk, or 'best', alternative restoration method (Table 7). The steps developed for TOPSIS technique application are the following:

- $\underline{Step\ 1}$: calculation of S $= \left\{ \sum_{i=1,\ j=1}^{i=1,\ j=m} Rv(i,j) \right\}^{1/2}$, where, i = number of criteria/sub-criteria, j = number of alternatives and Rv(i, j) the risk severity value of each alternative j over the criterion i (data extracted from Table 2) and division of each Rv(i, j) element by S to get the R(i, j) vector;

- $\underline{Step\ 2}$: Multiplication of each R(i, j) element by WR(i) to get the vector V(i, j);

- $\underline{Step\ 3}$: Determination of the ideal solution A* by forming the vector V*(j) that contains the minimum value elements of vector V(i, j) (the lower risk element of each criterion);

- $\underline{Step\ 4}$: Determination of the negative ideal solution A′ by forming the vector V′(j) that contains the maximum value elements of vector V(i, j) (i.e., the higher risk element of each criterion);

- $\underline{Step\ 5}$: Calculation of the separation from the ideal solution (Euclidean distance) by forming the vector Si* = $\{ \sum_{j=1}^{j=m} [V(j)^* - V(i,j)]^2 \}^{1/2}$;

- $\underline{Step\ 6}$: Calculation of the separation from the negative ideal solution (Euclidean distance) by forming the vector Si′= $\{ \sum_{j=1}^{j=m} [V(j)′ - V(i,j)]^2 \}^{1/2}$;

- $\underline{Step\ 7}$: Calculation of the relative closeness to the ideal solution Cj* = Sj′*(Sj* + Sj)$^{-1}$; the elements of vector Cj* represent the score of each alternative; the optimum, or 'best' or most advantageous alternative, is the one with the highest total score: max [Cj*] = max [C_1*, C2*, ... , C_j*].

The final ranking is: C3* = 0.819 < C2* = 0.515 < C1* = 0.447 (Table 11). The lower risk alternative with the higher performance in the scheme AHP/TOPSIS is A3, which refers to the combined restoration technology. The second best is A2 (restoration focused on natural processes) and the last in the sequence, the higher risk alternative, is A1 (technical restoration). This result clearly indicates that the evaluators (experts) consider the alternative A3 as optimal, taking into account that this has the lower total risk score and also that it is based on the balance between the technical activities and the natural processes at the mined-out area.

Table 11. Results of the Combined AHP and TOPSIS (technique for order of preference by similarity to ideal solution) Methodology—Final Ranking of Alternatives.

Risk Criteria/Subcriteria	Risk-ID	WR(i)	WC(i, j)	$Rv(1,j)^2$	$Rv(2,j)^2$	$Rv(3,j)^2$	$\{\Sigma X(i,j)^2\}^{1/2}$	$R(1,j)$	$R(2,j)$	$R(3,j)$	$V(1,j)$	$V(1,j)$	$V(1,j)$	Ideal Solution Ai''	Negative Ideal Solution Ai'	Separation from Ideal Solution $S(1,j)^*$	Separation from Ideal Solution $S(2,j)^*$	Separation from Ideal Solution $S(3,j)^*$	Separation from Negative Ideal Solution $S(1,j)'$	Separation from Negative Ideal Solution $S(2,j)'$	Separation from Negative Ideal Solution $S(3,j)'$
						AHP Results							TOPSIS Results								
R1	*WR1*	**0.110**																			
R1.1	*WR11*	0.180	0.020	16.000	1.000	4.000	4.583	0.873	0.218	0.436	0.017	0.004	0.009	0.004	0.017	0.000	0.000	0.000	0.000	0.000	0.000
R1.2	*WR12*	0.200	0.022	16.000	1.000	4.000	4.583	0.873	0.218	0.436	0.019	0.005	0.010	0.005	0.019	0.000	0.000	0.000	0.000	0.000	0.000
R1.3	*WR13*	0.180	0.020	4.000	1.000	4.000	3.000	0.667	0.333	0.667	0.013	0.007	0.013	0.007	0.013	0.000	0.000	0.000	0.000	0.000	0.000
R1.4	*WR14*	0.120	0.013	4.000	4.000	4.000	3.464	0.577	0.577	0.577	0.008	0.003	0.008	0.008	0.008	0.000	0.000	0.000	0.000	0.000	0.000
R1.5	*WR15*	0.100	0.011	16.000	1.000	4.000	4.583	0.873	0.218	0.436	0.010	0.002	0.005	0.002	0.010	0.000	0.000	0.000	0.000	0.000	0.000
R1.6	*WR16*	0.220	0.024	16.000	4.000	4.000	4.899	0.816	0.408	0.408	0.020	0.010	0.010	0.010	0.020	0.000	0.000	0.000	0.000	0.000	0.000
R2	*WR2*	**0.120**																			
R2.1	*WR2.1*	0.160	0.019	4.000	16.000	4.000	4.899	0.408	0.816	0.408	0.008	0.015	0.008	0.008	0.016	0.000	0.000	0.000	0.000	0.000	0.000
R2.2	*WR2.2*	0.200	0.024	4.000	16.000	4.000	4.899	0.408	0.816	0.408	0.010	0.020	0.010	0.010	0.020	0.000	0.000	0.000	0.000	0.000	0.000
R2.3	*WR2.3*	0.170	0.020	4.000	16.000	4.000	4.899	0.408	0.816	0.408	0.008	0.017	0.008	0.008	0.017	0.000	0.000	0.000	0.000	0.000	0.000
R2.4	*WR2.4*	0.140	0.017	16.000	4.000	4.000	4.899	0.816	0.408	0.408	0.014	0.007	0.007	0.007	0.014	0.000	0.000	0.000	0.000	0.000	0.000
R2.5	*WR2.5*	0.120	0.014	1.000	16.000	4.000	4.583	0.218	0.873	0.436	0.003	0.013	0.006	0.003	0.013	0.000	0.000	0.000	0.000	0.000	0.000
R2.6	*WR2.6*	0.210	0.025	4.000	1.000	1.000	2.449	0.816	0.408	0.408	0.021	0.010	0.010	0.010	0.021	0.000	0.000	0.000	0.000	0.000	0.000

Table 11. *Cont.*

	AHP Results							TOPSIS Results													
Risk Criteria/Subcriteria	Risk-ID	$WR(i)$	$WC(i,j)$	$Rv(1,j)^2$	$Rv(2,j)^2$	$Rv(3,j)^2$	$\{\Sigma X(\iota,j)^2\}^{1/2}$	$R(1,j)$	$R(2,j)$	$R(3,j)$	$V(1,j)$	$V(1,j)$	$V(1,j)$	Ideal Solution Ai^*	Negative Ideal Solution Ai'	Separation from Ideal Solution $S(1,j)^*$	Separation from Ideal Solution $S(2,j)^*$	Separation from Ideal Solution $S(3,j)^*$	Separation from Negative Ideal Solution $S(1,j)'$	Separation from Negative Ideal Solution $S(2,j)'$	Separation from Negative Ideal Solution $S(3,j)'$
R3	**WR3 0.161**																				
R31	WR3.1 0.310	0.050	16.000	4.000	1.000	4.583	0.873	0.436	0.218	0.044	0.022	0.011	0.011	0.044	0.001	0.000	0.000	0.000	0.000	0.001
R32	WR3.2 0.180	0.029	16.000	4.000	1.000	4.583	0.873	0.436	0.218	0.025	0.013	0.006	0.006	0.025	0.000	0.000	0.000	0.000	0.000	0.000
R33	WR3.3 0.270	0.044	16.000	4.000	1.000	4.583	0.873	0.436	0.218	0.038	0.019	0.010	0.010	0.038	0.001	0.000	0.000	0.000	0.000	0.001
R34	WR3.4 0.240	0.039	4.000	16.000	4.000	4.899	0.408	0.816	0.408	0.016	0.032	0.016	0.016	0.032	0.000	0.000	0.000	0.000	0.000	0.000
R4	**WR4 0.349**																				
R4.1	WR4.1 0.220	0.077	16.000	1.000	1.000	4.243	0.943	0.236	0.236	0.072	0.018	0.018	0.018	0.072	0.003	0.000	0.000	0.000	0.003	0.003
R4.2	WR4.2 0.075	0.026	16.000	4.000	1.000	4.583	0.873	0.436	0.218	0.023	0.011	0.006	0.006	0.023	0.000	0.000	0.000	0.000	0.000	0.000
R4.3	WR4.3 0.075	0.026	1.000	16.000	4.000	4.583	0.218	0.873	0.436	0.006	0.023	0.011	0.006	0.023	0.000	0.000	0.000	0.000	0.000	0.000
R4.4	WR4.4 0.033	0.012	16.000	1.000	4.000	4.583	0.873	0.218	0.436	0.010	0.003	0.005	0.003	0.010	0.000	0.000	0.000	0.000	0.000	0.000
R4.5	WR4.5 0.090	0.031	1.000	16.000	4.000	4.583	0.218	0.873	0.436	0.007	0.027	0.014	0.007	0.027	0.000	0.000	0.000	0.000	0.000	0.000
R4.6	WR4.6 0.140	0.049	4.000	16.000	4.000	4.899	0.408	0.816	0.408	0.020	0.040	0.020	0.020	0.040	0.000	0.000	0.000	0.000	0.000	0.000
R4.7	WR4.7 0.098	0.034	1.000	16.000	4.000	4.583	0.218	0.873	0.436	0.007	0.030	0.015	0.007	0.030	0.000	0.001	0.000	0.001	0.000	0.000
R4.8	WR4.8 0.075	0.026	16.000	1.000	4.000	4.583	0.873	0.218	0.436	0.023	0.006	0.011	0.006	0.023	0.000	0.000	0.000	0.000	0.000	0.000
R4.9	WR4.9 0.085	0.030	1.000	16.000	4.000	4.583	0.218	0.873	0.436	0.006	0.026	0.013	0.006	0.026	0.000	0.000	0.000	0.000	0.000	0.000
R4.10	WR4.10 0.062	0.022	4.000	1.000	4.000	3.000	0.667	0.333	0.667	0.014	0.007	0.014	0.007	0.014	0.000	0.000	0.000	0.000	0.000	0.000
R4.11	WR4.11 0.047	0.016	4.000	1.000	1.000	2.449	0.816	0.408	0.408	0.013	0.007	0.007	0.007	0.013	0.000	0.000	0.000	0.000	0.000	0.000

Table 11. *Cont*

Risk Criteria/Subcriteria	Risk-ID	AHP Results $WR(i)$	$WC(i, j)$	$Rv(1, j)^2$	$Rv(2, j)^2$	$Rv(3, j)^2$	$\{\Sigma X(\iota, j)^2\}^{1/2}$	$R(1,j)$	$R(2,j)$	$R(3,j)$	$V(1,j)$	$V(1,j)$	$V(1,j)$	Ideal Solution Ai*	Negative Ideal Solution Ai'	TOPSIS Results Separation from Ideal Solution $S(1,j)$*	Separation from Ideal Solution $S(2,j)$*	Separation from Ideal Solution $S(3,j)$*	Separation from Negative Ideal Solution $S(1,j)$'	Separation from Negative Ideal Solution $S(2,j)$'	Separation from Negative Ideal Solution $S(3,j)$'
R5	**WR5**	**0.260**																			
R5.1	WR5.1	0.010	0.003	4.000	4.000	4.000	3.464	0.577	0.577	0.577	0.001	0.001	0.001	0.001	0.001	0.000	0.000	0.000	0.000	0.000	0.000
R5.2	WR5.2	0.128	0.033	1.000	16.000	4.000	4.583	0.218	0.873	0.436	0.007	0.029	0.015	0.007	0.029	0.000	0.000	0.000	0.000	0.000	0.000
R5.3	WR5.3	0.112	0.029	1.000	4.000	1.000	2.449	0.408	0.816	0.408	0.012	0.024	0.012	0.012	0.024	0.000	0.000	0.000	0.000	0.000	0.000
R5.4	WR5.4	0.147	0.038	1.000	4.000	1.000	2.449	0.408	0.816	0.408	0.016	0.031	0.016	0.016	0.031	0.000	0.000	0.000	0.000	0.000	0.000
R5.5	WR5.5	0.091	0.024	1.000	16.000	4.000	4.583	0.218	0.873	0.436	0.005	0.021	0.010	0.005	0.021	0.000	0.000	0.000	0.000	0.000	0.000
R5.6	WR5.6	0.099	0.026	4.000	16.000	4.000	4.899	0.408	0.816	0.408	0.010	0.021	0.010	0.010	0.021	0.000	0.000	0.000	0.000	0.000	0.000
R5.7	WR5.7	0.079	0.021	4.000	16.000	4.000	4.899	0.408	0.816	0.408	0.008	0.017	0.008	0.008	0.017	0.000	0.000	0.000	0.000	0.000	0.000
R5.8	WR5.8	0.098	0.025	1.000	16.000	1.000	4.243	0.236	0.943	0.236	0.006	0.024	0.006	0.006	0.024	0.000	0.000	0.000	0.000	0.000	0.000
R5.9	WR5.9	0.039	0.010	4.000	16.000	1.000	4.583	0.436	0.873	0.218	0.004	0.009	0.002	0.002	0.009	0.000	0.000	0.000	0.000	0.000	0.000
R5.10	WR5.10	0.057	0.015	4.000	16.000	4.000	4.899	0.408	0.816	0.408	0.006	0.012	0.006	0.006	0.012	0.000	0.000	0.000	0.000	0.000	0.000
R5.11	WR5.11	0.076	0.020	1.000	4.000	1.000	2.449	0.408	0.816	0.408	0.008	0.016	0.008	0.008	0.016	0.000	0.000	0.000	0.000	0.000	0.000
R5.12	WR5.12	0.019	0.005	4.000	4.000	4.000	3.464	0.577	0.577	0.577	0.003	0.003	0.003	0.003	0.003	0.000	0.000	0.000	0.000	0.000	0.000
R5.13	WR5.13	0.045	0.012	1.000	16.000	1.000	4.243	0.236	0.943	0.236	0.003	0.011	0.003	0.003	0.011	0.000	0.000	0.000	0.000	0.000	0.000
																0.082	**0.068**	**0.021**	**0.066**	**0.072**	**0.096**

Table 11. *Cont.*

Alternative A1	$S1' = 0.066$	$S2^* = 0.082$	$S1^*+S1' = 0.148$	$C1^*= S1'/(S1^*+S1') \sim$ **0.447**
Alternative A2	$S2' = 0.072$	$S2^* = 0.068$	$S2^*+S2' = 0.141$	$C2^*= S2'/(S2^*+S2') \sim$ **0.515**
Alternative A3	$S3' = 0.096$	$S3^* = 0.021$	$S2^*+S2' = 0.117$	$C3^*= S3'/(S3^*+S3') \sim$ **0.819**
Final Ranking of Alternatives: A3 > A2 > A1				**Alternative A3 is the Lower Risk Restoration method**

5.6. Discussion

The analysis confirms that the suggested methodology can provide a decision making framework for the landscape planning and restoration of surface mining projects. In this context, the proposed approach (a) ensures objectivity in the identification and determination of the relative weight of each risk factor/sub-factor and (b) demonstrates the selection of the optimum restoration technology in a clear and explicit way, enabling decision makers to understand the results of the multi-criteria evaluation process and to make a reasonable decision.

The lack of precise quantitative data in the mining projects makes the complete understanding of the restoration needs and perspectives quite difficult. To overcome this deficiency, the study suggests the knowledge elicitation of mining managers and production engineers and the transformation in an explicit form by combining AHP and TOPSIS techniques.

The combination of the two techniques allows better decision making, considering the advantages and the shortcomings of each technique. For example, since TOPSIS by its nature (based on the Euclidean distance) does not allow correlation of the criteria, the correlation can be achieved by means of the AHP pairwise comparison.

Regarding the strengths of the AHP/TOPSIS combination, the AHP is flexible in application and allows the consideration of subjective and objective factors in the decision making process. Furthermore, in AHP applications, the experts, through interpersonal contacts or properly prepared questionnaires, offer their judgment for the structuring and processing of reciprocal matrices, where the risk factors are identified and correlated. The higher experience of the experts in the management of mining restoration risks ensures the better identification and evaluation of the relative weights of risk factors and sub-factors. In this work, the option of interviewing three qualified experts with long term experience (>20 years) in surface mining systems was adopted. In this way, the relative weights of the identified risks meet a high level of technical justification and reliability that increases the objectivity of the technique, to the extent possible. In addition, the application of TOPSIS allows the "bounding" of the risk ranking between the "positive ideal" and the "negative ideal" solution, thus, the definition of the total risk is combined with mathematical clarity and simplicity.

On the other hand, with the combination of the AHP and TOPSIS techniques, the subjectivity cannot be eliminated, since the experts have their own perceptions and biases on the significance of each individual risk in relation to the restoration alternatives. This problem is more complicated in cases where the group of experts is not exclusively composed of mining specialists, but also includes representatives assigned by municipalities, public agencies, NGOs, etc., with different educational background, environmental perception or objectives. In such cases, many repetitions of the AHP may be required to fulfill the consistency criterion. Another weakness is the performance limitation and inefficiency, especially in cases where the number of alternatives and/or the number of risk factors and sub-factors is high.

In conclusion, for the successful implementation of the methodology, an appropriate organization and control of the whole decision making context is required. The objectives of the restoration project should be well defined, and the weaknesses and inefficiencies of the AHP/TOPSIS applications must be identified and considered.

6. Conclusions and Perspectives for Further Research

The MCDM philosophy could be adopted to support the risk management needs in projects of surface lignite mines restoration. The selection of a lower risk restoration technology/method is critical as it provides a substantial basis for the decision makers to realize that a restoration project is technically feasible and environmentally friendly, with benefits for the environment, society and economy, and can be completed meeting the cost and time constraints.

This work makes an important contribution to the field of geo-energy, by demonstrating how the principles of risk management can be adapted in complex projects of mining restoration on the basis of computational simplicity of the applied MCDM methods. In this context, the appropriate multi-criteria methodology can be an efficient decision making support tool in terms of objectivity, mathematical consistency and clarity of the quantitative results.

The suggested methodology enables the evaluation and selection of technically appropriate and low risk technologies/methods for the restoration/reclamation of closing surface mines through the combination of AHP and TOPSIS techniques, with an aim to solve practical risk-based decision making problems. However, the combined AHP and TOPSIS application has specific limitations, especially in cases where the number of criteria or the alternatives is high, or the group of experts presents educational, professional or cultural heterogeneities.

The AHP method, as a decision-making tool, is simple and easy to use and allows the low cost analysis and quantification of project risks, by aggregating the knowledge of mining and restoration experts and transforming it in an explicit form. The TOPSIS uses the relative weights of the risk-based criteria and sub-criteria, identified by the AHP, as inputs of the numerical calculations for the definition of the ideal and negatively ideal solutions and, therefore, for obtaining the ranking of mine restoration alternatives. Both techniques are used, on an individual basis or in combination with others, in the evaluation of mining projects.

Finally, some proposals for further research are pointed out. One relates to the investigation of restoration risks in a more analytical concept. For example, the crisp values 1, 2 and 4 corresponding to risk impacts classification as low, medium and high, could be further analyzed in five levels, by defining interim numerical values between low and medium and also between medium and high levels of severity. Thus, the classification of risk impacts can be analyzed in a more detailed basis. Moreover, the aforementioned crisp values can be interpreted in a form of fuzzy variables to express the uncertainty of each specific risk using simplified linguistic expressions. Therefore, the methodology could be further improved in more sophisticated approaches. Moreover, other MCDM techniques such as PROMETHEE (I or II), ELECTRE, SMART, VIKOR, DEMATEL, fuzzy versions of these techniques, Bayesian networks, neuro-fuzzy algorithms or other hybrid techniques could be used instead of TOPSIS for risks quantification and processing. Finally, a comparative analysis of the performance and efficiency of various MCDM methods in a decision making framework related to mine restorations can be performed.

Author Contributions: Conceptualization, P.-M.S., C.R. and F.P.; methodology, P.-M.S., C.R. and F.P.; validation, P.-M.S., C.R. and F.P.; formal analysis, P.-M.S., C.R. and F.P.; data curation, P.-M.S., C.R. and F.P.; writing—original draft preparation, P.-M.S., C.R. and F.P.; writing—review and editing, P.-M.S., C.R. and F.P.; visualization, P.-M.S., C.R. and F.P.; supervision, P.-M.S., C.R. and F.P.; project administration, P.-M.S., C.R. and F.P. All authors have read and agreed to the published version of the manuscript.

Funding: This research received no external funding.

Acknowledgments: The authors would like to thank Professor Zach Agioutantis, Department of Mining Engineering, University of Kentucky, for his constructive comments on this paper.

Conflicts of Interest: The authors declare no conflict of interest.

References

1. Schulz, F.; Wiegleb, G. Development Options of natural habitats in a post-mining landscape. *Land Degrad. Dev.* **2000**, *11*, 99–110. [CrossRef]
2. Chuman, T. Restoration Practices used on Post mining Sites and Industrial Deposits in the Czech Republic with an Example of Natural Restoration of Granodiorite Quarries and Spoil Heaps. *J. Landsc. Ecol.* **2015**, *8*, 29–46. [CrossRef]
3. Prach, K.; Hobbs, R.J. Spontaneous succession versus technical reclamation in the restoration of disturbed sites. *Restor. Ecol.* **2008**, *16*, 363–366. [CrossRef]
4. Tischew, S.; Krimer, A.; Lorenz, A. Alternative Restoration Strategies in Former Lignite Mining Areas of Eastern Germany. In *Biodiversity: Structure and Function*, 2nd ed.; EOLSS Publishers Co. Ltd: Oxford, UK, 2009.

5. Sinha, S. *Overview on Reclamation and rehabilitation of Mines*; Indian Mineral Sector Indian Bureau of Mines: Rajastan, India, 2015.

6. Imboden, C.; Moczek, N. Risks and opportunities in the biodiversity management and related stakeholder involvement of the RWE Hambach Lignite Mine. *Int. Union Conserv. Nat.* **2015**. Available online: www.iucn.org/publications (accessed on 15 January 2020).

7. Roumpos, C.; Pavloudakis, F.; Galetakis, M. Modelling and evaluation of open-pit lignite mines exploitation strategy. In Proceedings of the 2nd International Conference on Sustainable Development Indicators in the Minerals Industry (SDIMI 2005), Aachen, Germany, 18–20 May 2005; pp. 1127–1139.

8. Paraskevis, N.; Roumpos, C.; Stathopoulos, N.; Adam, A. Spatial analysis and evaluation of a coal deposit by coupling AHP & GIS techniques. *Int. J. Min. Sci. Technol.* **2019**, *29*, 943–953.

9. Pavloudakis, F.; Galetakis, M.; Roumpos, C. A spatial decision support system for the optimal environmental reclamation of open-pit coal-mines in Greece. *Int. J. Min. Reclam. Environ.* **2009**, *23*, 291–303. [CrossRef]

10. Kim, S.M.; Choi, Y.; Suh, J.; Oh, S.; Park, H.D.; Yoon, S.H.; Goc, W.R. ArcMine: A GIS extension to support mine reclamation planning. *Comput. Geosci.* **2012**, *46*, 84–95. [CrossRef]

11. Palogos, I.; Galetakis, M.; Roumpos, C.; Pavloudakis, F. Selection of optimal land uses for the reclamation of surface mines by using evolutionary algorithms. *Int. J. Min. Sci. Technol.* **2017**, *27*, 491–498. [CrossRef]

12. Roumpos, C.; Spanidis, P.M.; Pavloudakis, F. Environmental Reclamation Planning of Continuous Surface Lignite Mines in Closure Phase: A Risk-Based Investigation. In Proceedings of the 14th International Symposium of Continuous Surface Mining, Thessaloniki, Greece, 23–26 September 2018; pp. 551–562.

13. Pavloudakis, F.; Spanidis, P.M.; Roumpos, C. Investigation of Natural and Technological Hazards and associated Risks in Continuous Surface Lignite Mines. In Proceedings of the 2nd International Conference on Natural Hazards & Infrastructure, Chania, Greece, 23–26 June 2019.

14. Wang, Y. Application of TOPSIS and AHP in the Multi-Objective Decision Making Problems. *Matec Web Conf.* **2018**, *228*, 05002. [CrossRef]

15. Dey, P.K. Project Risk Management: A Combined Analytic Hierarchy Process and Decision Tree Approach. *Cost Eng. March* **2002**, *44*, 13–26.

16. Dey, P.K. Project risk management using multiple criteria decision-making technique and decision tree analysis: A case study of Indian oil refinery. *Prod. Plan. Control* **2012**, *23*, 903–921. [CrossRef]

17. Emrouznejad, A.; Marra, M. The state of the art development of AHP (1979–2017): A literature review with a social network analysis. *Int. J. Prod. Res.* **2017**, *55*, 6653–6675. [CrossRef]

18. Mahase, M.J.; Musingwini, C.; Nhleko, A.S. A survey of applications of multi-criteria decision analysis methods in mine planning and related case studies. *J. S. Afr. Inst. Min. Metall.* **2016**, *116*, 1051–1056. [CrossRef]

19. Sousa, T.W., Jr.; Sousa, M.J.F.; Cabral, I.E.; Diniz, M.E. Multi-criteria decision making aid methodology applied to highway truck selection at a mining company. Revista Escola de Minas. *Ouro Preto* **2014**, *67*, 285–290.

20. Musingwini, C. A review of the theory and application of multi-criteria decision making analysis techniques in mine planning. In *Proceedings of Mine Planning and Equipment Selection (MPES)*; Australasian Institute of Mining and Metallurgy: Melbourne, Australia, 2010; pp. 129–140.

21. Musingwini, C.; Minnitt, R.A. Ranking the efficiency of selected platinum mining methods using the analytical hierarchy process. In Proceedings of the Third International Platinum Conference, Johannesburg, South Africa, 6–9 October 2008; pp. 319–326.

22. Bitarafan, M.R.; Ataei, M. Mining method selection by multiple criteria decision making tools. *J. S. Afr. Inst. Min. Metall.* **2004**, *104*, 493–498.

23. De Almeida, A.T.; Alencar, L.H.; De Miranda, C.M.G. Mining methods selection based on multi-criteria models. In *Proceedings of the Application of Computers and Operations Research in the Mineral Industry*; Taylor and Francis Group: London, UK, 2005; Volume 449, pp. 19–24.

24. Yavouz, M. Equipment selection based on the AHP and Yager's method. *J. S. Afr. Inst. Min. Metall.* **2015**, *115*, 425–433. [CrossRef]

25. Karadogan, A.; Kahriman, A.; Ozer, U. Application of fuzzy set theory in the selection of an underground mining method. *J. S. Afr. Inst. Min. Metall.* **2015**, *108*, 73–79.

26. Kazakidis, V.N.; Mayer, Z.; Scoble, M.J. Decision making using the analytic hierarchy process in mining engineering. *Min. Technol.* **2004**, *113*, 30–42. [CrossRef]

27. Khakestar, M.S.; Hassani, H.; Moarefvand, P.; Madani, H. Application of multi-criteria decision making methods in slope stability analysis of open pit mines. *J. Geol. Soc. India* **2016**, *87*, 213–221. [CrossRef]

28. Bazzazi, A.A.; Osanloo, M.; Karimi, B. Optimal Open Pit Lignite Mining Equipment Selection using Fuzzy Multiple Attribute Decision Making approach. *Arch. Min. Sci.* **2009**, *54*, 301–320.

29. Bascetin, A. A decision support system using analytical hierarchy process (AHP) for the optimal environmental reclamation of an open-pit mine. *Environ. Geol.* **2007**, *52*, 663–672. [CrossRef]

30. Uberman, R.; Ostrega, A. Applying the analytic hierarchy process in the revitalization of post-mining areas field. In Proceedings of the ISAHP' 05: 8th International Symposium on the Analytic Hierarchy Process (ISAHP), Honolulu, HI, USA, 8–10 July 2005.

31. Sitorus, F.; Cilliers, J.J.; Brito-Parada, P.R. Multi-criteria decision making for the choice problem in mining and mineral processing: Applications and trends. *Expert Syst. Appl.* **2019**, *121*, 393–417. [CrossRef]

32. Velasquez, M.; Hester, P.T. An Analysis of Multi-Criteria Decision Making Methods. *Int. J. Oper. Res.* **2013**, *10*, 56–66.

33. Krohling, R.A.; De Souza, T.T.M. Two Examples of Application of TOPSIS to Decision Making Problems. *Rev. Sist. Inf.* **2011**, *8*, 31–35.

34. Ghanbarpour, M.R.; Hipel, K.W. Multi-Criteria Planning Approach for Ranking of Land Management Alternatives at Different Scales. *Res. J. Environ. Earth Sci.* **2011**, *2*, 167–176.

35. Parida, P.K.; Sahoo, S.K. Multiple Attributes Decision Making Approach by TOPSIS Technique. *Int. J. Eng. Res. Technol.* **2013**, *2*, 907–912.

36. Saaty, T.L. *The Analytic Hierarchy Process*; McGraw Hill International: New York, NY, USA, 1980.

37. Al-Harbi, K.A. Application of the AHP in project management. *Int. J. Proj. Manag.* **2001**, *19*, 19–27. [CrossRef]

38. Saaty, T.L. Decision making with the Analytic Hierarchy Process. *Int. J. Serv. Sci.* **2008**, *1*, 83–98. [CrossRef]

39. Bradshaw, A. Restoration of mined lands using natural processes. *Ecol. Eng.* **1997**, *8*, 255–269. [CrossRef]

40. LUMINANT. *An Overview of Lignite Mine Reforestation at Luminant's Martin Lake Mines in Eastern Texas*; Luminant Co: Dallas, TX, USA, 2015.

41. Pavloudakis, F.; Agioutantis, Z. Using environmental permits for boosting the environmental performance of large-scale lignite surface mining activities in Greece. In Proceedings of the 2008 National Meeting of the American Society of Mining and Reclamation, Richmond, VA, USA, 14–19 June 2008.

42. Kasztelewicz, Z. Approaches to post-mining land reclamation in Polish open-cast lignite mining. *Civ. Environ. Eng. Rep.* **2014**, *12*, 55–67. [CrossRef]

43. Erener, A. Remote sensing of vegetation health for reclaimed areas of Seyitomer open cast coal mine. *Int. J. Coal Geol.* **2011**, *86*, 20–26. [CrossRef]

44. Triantaphyllou, E. *Multi-Criteria Decision Making Methods: A Comparative Study*; Kluwer Academic: Boston, MA, USA, 2000.

45. Saaty, T.L.; Vargas, L.G. *Projection and Forecasting*; Kluwer Academic Publishers: Dordrecht, The Netherlands, 1991; p. 251.

46. Sloss, L. *Coal Mine Site Reclamation*; CCC/216; IEA Clean Coal Centre: London, UK, 2013; ISBN 978-92-9029-536-5.

47. Leathers, K. *Costs of Strip Mine Reclamation in the West*; US Department of Agriculture, Rural Development Research Report; No.19; Economics, Statistics and Cooperative Service, Department of Agriculture: Washington, DC, USA, 1980.

48. Rovolis, A.; Kalimeris, P. Roadmap for the Transition of the Western Macedonia Region to a Post Lignite Era. WWF, Economic and Technical Assessment. 2016. Available online: www.wwf.gr/images/pdfs/Rmap_Study.pdf (accessed on 1 March 2020).

49. Knabe, W. Methods and Results of Trip-Mine Reclamation in Germany. *Ohio J. Sci.* **1964**, *64*, 74–105.

50. PMI. *A Guide to Project Management Body of Knowledge (PMBOK)*, 6th ed.; Project Management Institute, Inc: Newtown Square, PA, USA, 2019.

51. Kirytopoulos, K. *Manual for Projects' Risk Management*; Klidarithmos Publications: Athens, Greece, 2010. (In Greek)

Article

The Time Duration of the Effects of Total Extraction Mining Methods on Surface Movement

André Vervoort

Department of Civil Engineering, KU Leuven, 3001 Leuven, Belgium; andre.vervoort@kuleuven.be;
Tel.: +32-16-321171

Received: 16 July 2020; Accepted: 6 August 2020; Published: 8 August 2020

Abstract: Since the 1990s, remote sensing data have been available to monitor the surface movement for long periods of time. The analysis of satellite data shows that there is still residual subsidence (i.e., with average rates of about −10 mm/year) several decades after mining longwall panels in an area. Several years after the underground infrastructure was sealed, the surface started to move upwards. In the past, it often was claimed that movement of the surface was limited in time, i.e., a few years after mining a longwall panel. This is not the case for the conditions of the Campine coal basin, Belgium. This knowledge is important when one wants to design new operations in deep coal seams, but also when planning to stop the underground mining and to seal the access to the mine.

Keywords: longwall coal mining; ground control; subsidence; uplift; surface movement; radar-interferometry; long-term behavior; sustainable mining; Belgium

---

1. Introduction

There is general agreement that the total extraction mining method, such as longwall mining with the associated goaf, has an impact on the environment, e.g., the subsidence of the surface [1–5]. It also is obvious that this impact is permanent, but that issue is not the focus of this paper. The problem being addressed herein is whether the surface movement will be an ongoing phenomenon long after all mining activities have stopped or whether the movement will reach a state of complete, stabilized equilibrium after a certain period. Generally, the people in the industry, as well as researchers, have the opinion that the surface movement is limited in time. Often, it is claimed that subsidence only occurs over a period of a few months to a few years and that, subsequently, the ground is stable. Some researchers divide the process of subsidence into various phases [6–9]. They identify the first phase as initial subsidence and the second phase that is referred to as either principal, accelerated, or steady-state subsidence. The third phase is referred to as residual, differed, or delayed subsidence, and it is assumed to be limited in time. After the third phase, it is assumed that the subsidence has stopped and that the ground is stable [7]. Some examples of time periods for the residual phase from various places in the world are given by Al Heib et al. [7], and their conclusion is as follows: *"The duration of the residual subsidence phase is about 12 to 18 months, but this duration is often less long, i.e., about 3 to 4 months when the exploitation is carried out in an already disturbed zone (several seams, goaf, …). There are some isolated cases resulting from geological contexts and/or particular exploitation, for these cases, the duration of residual subsidence can appreciably be raised and spread out over one period of 4 to 6 years."* The end of the residual subsidence phase is defined as either (1) when an amount of additional subsidence corresponding to a certain percentage of the subsidence during the two previous phases has been reached, (2) when the increments between successive measurements have become negligible, or (3) when a predefined fixed time period has been reached.

The basis for conducting the research presented in this paper was not a specific interest in long-term subsidence; rather, it was the occurrence of partly unexpected and new behavior of the surface movement several years after the closure of the last coal mines in northeastern Belgium around 1990 [10,11]. A few years after the closure of the underground infrastructure and the dismantling of the underground pumping stations, the surface started to move upward. This phenomenon first was observed in the Netherlands, where the coal mines were closed in the 1970's [12]. Efforts to understand and explain this uplift phenomenon led to the study of the residual subsidence prior to the start of uplift, and hence, to the study of the long-term impact of deep coal mining.

The Campine basin belongs to the Upper Carboniferous strata (Westphalian stage). It is part of the South Permian basin of northwestern Europe [10]. Mainly coking or metallurgical coal was mined, with an average calorific value of 35 MJ/kg. The overburden contains sand, clay and chalk; i.e., all weak geological material. The total thickness of the overburden varies between 400 and 600 m. It contains several aquifers and aquitards [13]. The waste rock between the mined coal seams is composed mainly of shale, siltstone, sandstone, and thin (unmined) coal layers. Overall, the successive layers of waste rock are relatively thin (on the order of decimeter to meter in scale).

2. Global View of Surface Movement after the Closure of Mines

This paper presents data for the western part of the Belgian Campine basin in northeastern Belgium. Three coal mines were active in that part of the basin, i.e., from west to east, the Beringen, Zolder, and Houthalen mines [10]. The depths of the mining in the western part of the Belgian Campine basin varied from −485 m to −967 m. The mining height for a single longwall varied from 0.5 m to 3.3 m, with an average of 1.5 m. In most areas, 5 to 10 coal seams were mined above each other. Coal production started for the three mines between 1922 and 1939, and all of the production was terminated between 1990 and 1992. The underground pumping stations were dismantled, and the shafts were closed in the years after the production was terminated. This western part of the Campine basin was characterized by a period of additional residual subsidence that was followed by uplift. An earlier study showed that a residual downward movement occurred for the Houthalen coal mine until the end of 1999, which was followed by limited surface movement from 2000 through 2004, and a clear uplift since 2005 [13]. However, spatial variations were observed.

Taking the timing of closure and the subsequent periods of residual subsidence and uplift into account, the images of the European C-band ERS1/2 and ENVISAT-ASAR satellites were the most suitable. These data were acquired for research through a European Space Agency (ESA) research proposal [14]. The periods that were recorded for the two sets were from August 1992 through December 2000 (87 cycles of 35 days) and from December 2003 through October 2010 (72 cycles of 35 days), respectively. Radar interferometry or Interferometry with Synthetic Aperture Radar (InSAR) is very suitable for studying large time series over large surface areas. The movements of reflective surfaces (i.e., the so-called "permanent scatterers" or "reflectors") were followed during all successive cycles of the satellite. This allowed the acquisition of significant spatial coverage of the areas that were studied, at least if the areas were part of a built environment.

Figure 1 presents a global view of the surface movement in the period of mainly residual subsidence (Figure 1a), and in the period of mainly uplift (Figure 1b), for the entire surface area above and around the three coal mines. For comparison purposes, the surface movement was extrapolated to a total time period of 10 years for each Figure. Hence, Figure 1a covers the period from August 1992 through August 2002, and Figure 1b covers the period from March 2003 through March 2013. As the variation as a function of time is not always linear (see further), the increase of surface subsidence over the last 20 months in the first period was added to the value recorded for December 2000, while for the second period the increase in uplift during the first nine months and the last 29 months were added. Since 1 mm/year generally is accepted as a possible error on the trend of surface movement recorded by these satellites [15,16], only reflectors with a minimum additional movement of 10 mm over 10 years are presented. The outside borders of the zone of the longwall panels that were mined are presented by straight blue lines. The data clearly show that the reflectors with a significantly larger movement occurred mainly within these borders or in the immediate vicinity.

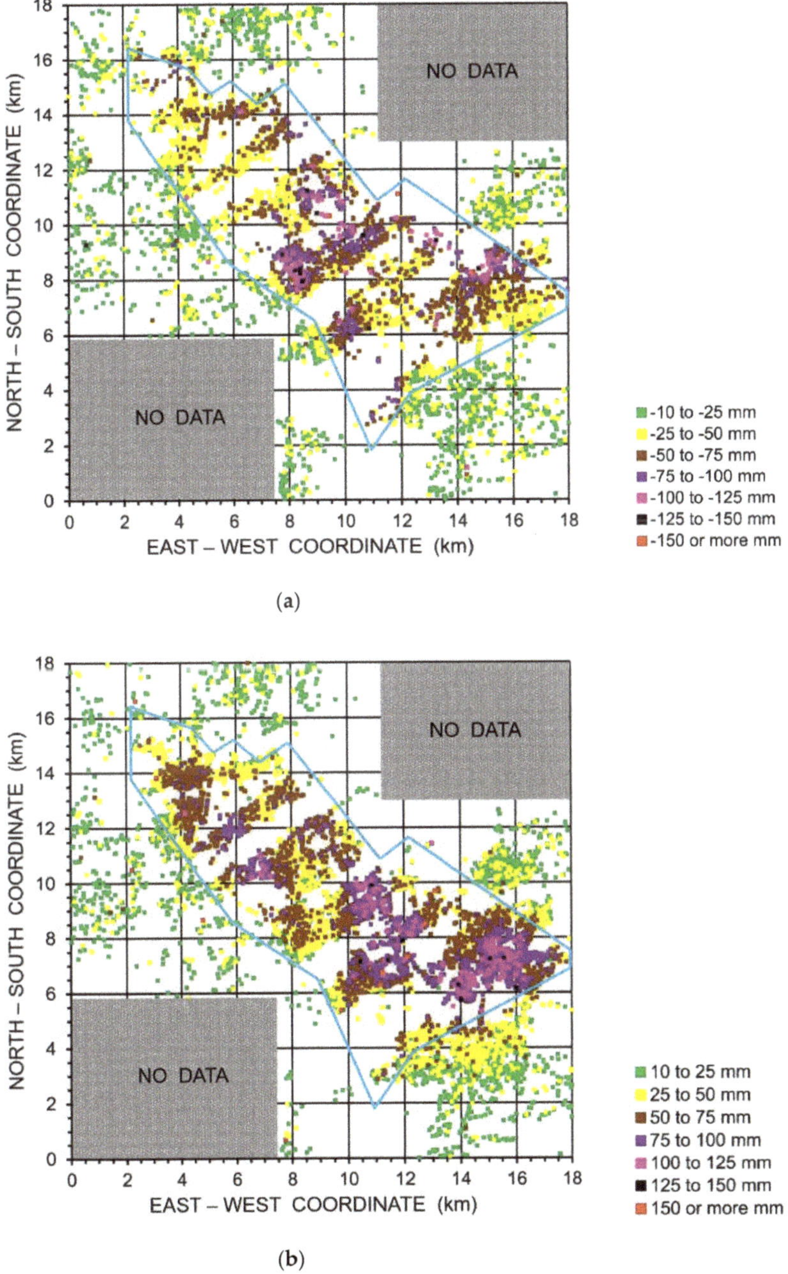

(a)

(b)

Figure 1. Spatial distribution of additional surface movement over a period of 10 years: (**a**) Period from August 1992 through August 2002, residual subsidence larger than −10 mm; (**b**) period from March 2003 through March 2013, uplift larger than 10 mm. The outside borders of the zone of the mined longwall panels are shown with the blue color.

However, some (smaller) movements are also recorded outside these borders. For a single panel, the angle of draw in the Campine basin is often assumed to be 45° [5]. In other words, the size of the zone of influence is approximately the depth of mining. Recent research has shown that this is an underestimation when looking at the entire coal basin. In Reference [5], this is illustrated by some examples: At about 3 km further than the boundary a systematic surface movement was recorded over a period of 20 years, while at a distance of 7 km no movement at all was recorded. As discussed later, by analyzing long-term frequent data series over large areas, new insights occur. With conventional levelling campaigns, one has the tendency to limit the area of monitoring and to stop monitoring once the difference in surface movement between successive levelling campaigns becomes equal to the accuracy of the levelling method.

Figure 2 presents the reflectors with an additional surface movement within ±10 mm over 10 years, as well as the reflectors with the opposite movement from the global trend (i.e., already uplift in the first period and still residual subsidence in the second period). In comparison to Figure 1, these two groups of reflectors now are situated mainly outside the mined area. In other words, one can conclude definitively that both types of movements (downward and upward) are linked to mining. The other conclusion is that two decades after the closure of the mine, there is still movement of the surface, and one cannot assume that the strata are stabilized.

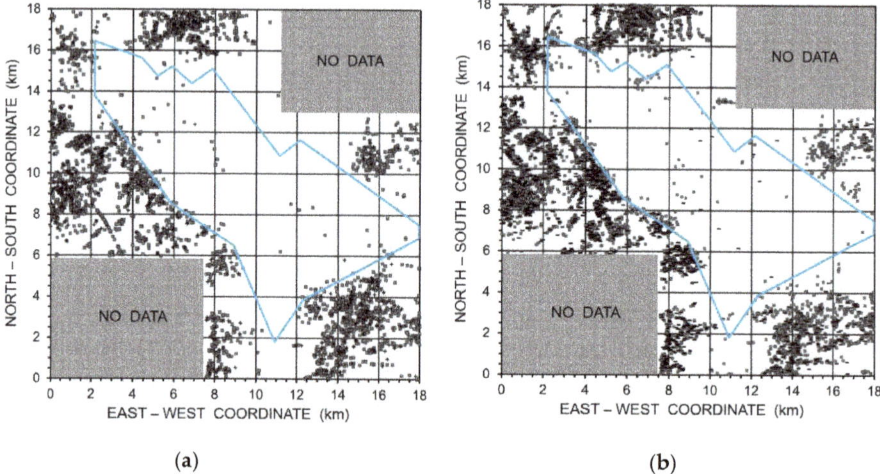

Figure 2. Spatial distribution of the reflectors with an additional surface movement of ±10 mm over 10 years (grey squares), as well as the reflectors with the movement that is opposite of the global trend (plus and minus signs, respectively): (**a**) Period from August 1992 through August 2002; (**b**) period from March 2003 through March 2013.

Globally, about one and a half to two times more reflectors were detected with the more modern second satellite (ENVISAT-ASAR) than with the first (European C-band ERS1/2). Some sub-areas within the mined area do not have any reflectors, e.g., the band between the east-west, north-south coordinates (4.5 km, 10.5 km) and (9.0 km, 13.0 km). These gaps where there are no reflectors in Figure 1 correspond to zones composed of natural land, forest, lakes, and agricultural land. Also, the area to the northeast of the mined area is such land.

When comparing in detail the spatial distribution of the surface movements between Figure 1a,b, one cannot conclude that the sub-areas with large residual subsidence correspond with sub-areas of large uplift. For example, in the northwest, the large uplift values for coordinates (6, 12) in Figure 1b (uplift values of mainly 75 to 100 mm over 10 years) correspond mainly to residual subsidence values of −25 to −75 mm in Figure 1a. Something similar can be observed for the entire western part in the northwest sub-area, i.e., within the east-west distance of 3 to 5 km and north-south between 11 and 14.5 km. Another example is the entire zone between east-west 11 and 17 km and north-south between 6 and 7 km. This zone is characterized by large uplift values and relatively low residual subsidence values. However, the zone around coordinates (8, 8) is characterized by high residual subsidence values (mainly −100 to −150 mm) and low uplift values (25 to 75 mm). All of these observations are in agreement with earlier research results [11,17] that showed that the spatial distributions of the residual subsidence values and the uplift values were different, and no correlation was observed between these two parameters. This information is useful in trying to understand the mechanisms that cause an uplift in comparison to the mechanisms that cause subsidence, as will be discussed later.

3. Surface Movement as A Function of Time Since Mining the Last Longwall Panel

3.1. Residual Subsidence

Since the overall research is focused on the new phenomenon, i.e., the occurrence of uplift several years after the closure of the underground infrastructure, four locations were selected that had relatively high uplift values.

Table 1 gives the coordinates of these four locations, which are referred to as the black, blue, green, and red locations. A sub-area was defined around each location in which 10 reflectors were present in the first observation period, i.e., the period with residual subsidence. The number of 10 reflectors was considered a good compromise between being able to calculate a meaningful average and not having a too large effect of the spatial variation.

Table 1. Information on the five sub-areas selected for detailed analyses (Figures 3–5): Coordinates of the central point in sub-area; depth interval and mining period of all longwall panels within extended sub-area; number of reflectors and minimum, average, and maximum additional surface movement (average rate between brackets) for both observation periods.

Sub-Area	East-West (km)	North-South (km)	Depth Interval (m)	Mining Period	First Period (8.37 year) N°	First Period (8.37 year) Min / Av / Max (mm (mm/year))	Second Period (6.92 year) N°	Second Period (6.92 year) Min / Av / Max (mm (mm/year))
Black	15.1	7.1	−560 … −966	1939 … 1973	10	−25.8 (−3.1) / −39.1 (−4.7) / −48.5 (−5.8)	20	35.3 (5.1) / 62.8 (9.1) / 80.6 (11.7)
Blue	10.8	9.1	−607 … −820	1932 … 1989	10	−31.9 (−3.8) / −54.0 (−6.5) / −65.2 (−7.8)	20	53.5 (7.7) / 68.2 (9.9) / 83.3 (12.0)
Green	5.8	11.9	−637 … −804	1933 … 1962	10	−27.7 (−3.3) / −36.7 (−4.4) / −51.7 (−6.2)	18	40.7 (5.9) / 55.2 (8.0) / 64.5 (9.3)
Red	8.8	12.1	−632 … −800	1968 … 1974	10	−53.8 (−6.4) / −60.6 (−7.2) / −76.2 (−9.1)	20	26.0 (3.8) / 42.6 (6.2) / 57.5 (8.3)
Brown	15.3	8.6	−609 … −967	1940 … 1992	10	−66.8 (−8.0) / −80.4 (−9.6) / −103.7 (−12.4)	20	25.4 (3.7) / 37.1 (5.4) / 45.2 (6.5)

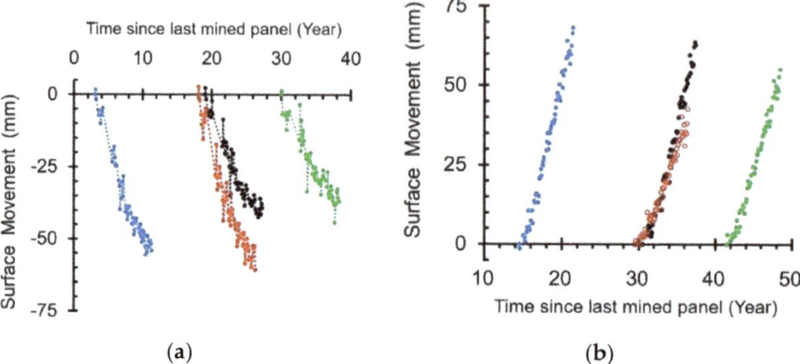

Figure 3. Variation of the average additional surface movement as a function of time since the last mined longwall panel for four sub-areas in the vicinity: (**a**) Period from August 1992 through December 2000; (**b**) period from December 2003 through October 2010.

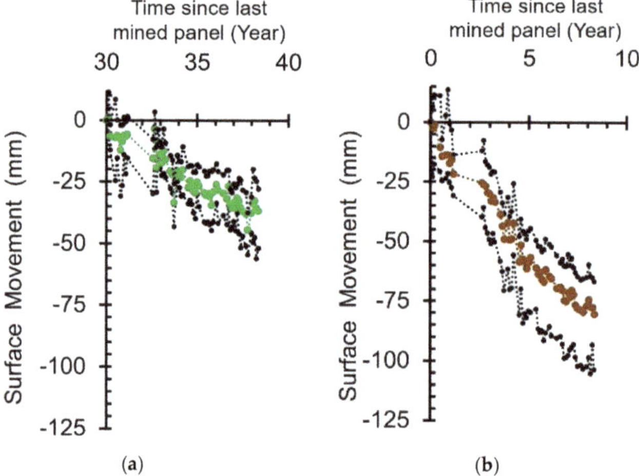

Figure 4. Variation of the minimum, average, and maximum residual subsidence as a function of time since the last mined longwall panel in the vicinity for two sub-areas over the period from August 1992 through December 2000: (**a**) Sub-area green; (**b**) sub-area brown.

This sub-area was extended further, reaching 1 km to the north, south, east, and west. All of the longwall panels that intersected this extended sub-area were considered, and the year of the most recent panel was taken as the origin of the time scale (Figure 3). An extension of 1 km was considered an acceptable distance for the zone of influence because it is larger than the maximum depth of mining. Figure 3a shows the variations of the average residual subsidence for the four sub-areas from August 1992 through December 2000. In comparison to the two previous Figures, the data of the satellite images are not extrapolated in Figure 3.

For all four sub-areas, residual subsidence is still occurring. For the green sub-area, this movement is occurring even though it has been 40 years since mining was done in the vicinity. Figure 4a shows the variation of the minimum, average, and maximum residual subsidence for this green sub-area. Note that the curve of the maximum values does not necessarily correspond to a single reflector. The same is true for the minimum values. The differences between the extreme values are relatively low, i.e., the values vary from −27.7 to −51.7 mm over the period of 8.37 years with an average additional subsidence of −36.7 mm. The latter corresponds to an average annual rate of −4.4 mm/year.

The differences between the minimum and maximum annual rates over the entire period with the average are 1.1 and 1.8 mm/year, respectively.

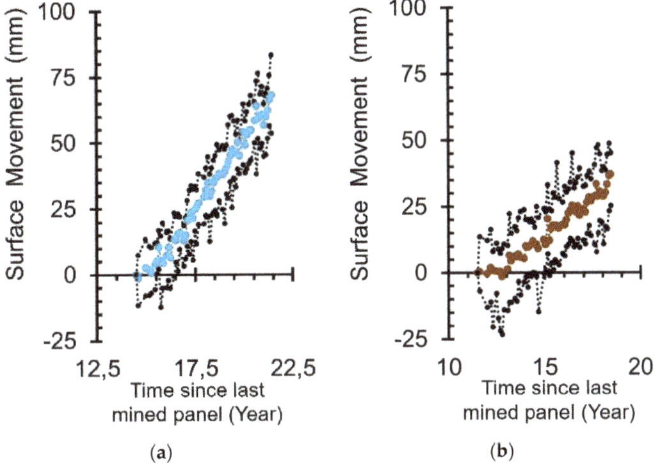

Figure 5. Variation of the minimum, average, and maximum uplift as a function of time since the last mined longwall panel in the vicinity for two sub-areas over the period from December 2003 through October 2010: (**a**) Sub-area blue; (**b**) sub-area brown.

Similar conclusions can be reached for the other sub-areas. (See Table 1 for the details.) The spread between the minimum and the maximum rates is a combination of both the inaccuracy of the measurement method and the impact of the spatial variation. Earlier research showed that the larger the study area is, the more variation there is. However, it was observed that by considering shorter time intervals, the larger the variation in the annual rates is. For an area with 117 reflectors, Vervoort and Declercq [13] estimated that the average rate over the first observation period (8.37 years) varied from −2.8 to −8.7 mm/year for these reflectors (so, per reflector, the average was calculated over the entire time period). However, the largest annual increase that was observed was −30 mm/year.

Figure 4b shows an example of the residual subsidence of a sub-area (brown) in which a longwall panel was mined just before the mine was closed. It also is a sub-area with relatively high residual subsidence values and lower uplift values. The central coordinates of this sub-area are (15.3, 8.6). The additional residual subsidence varies between −66.8 and −103.7 mm. The average annual rate over the entire time period is −9.6 mm/year. (The extremes are −8.0 and −12.4 mm/year, respectively.) So, the average annual rate is more than twice the rate of the green sub-area, which was the smallest of the four sub-areas of Figure 3a.

For the entire area presented in Figure 1a, the largest increase in residual subsidence was −148.1 mm over the entire observation period or an annual average of −17.7 mm/year (coordinates of 11.872 km east-west and 11.119 km north-south).

3.2. Uplift

For the same sub-areas in Figure 3a, Figure 3b shows the average variation of the uplift for the satellite images from December 2003 through October 2010. Therefore, each curve began in December 2003, but the origin of the time is the same as it was for Figure 3a, i.e., when the last panel was mined within the extended sub-area, and the same size of the sub-area was considered. Hence, the number of reflectors on which the average is based is larger than in Figure 3a (see Table 1), i.e., 18 to 20 reflectors rather than the fixed number of 10.

Figure 3b shows that the surface is still moving upwards about 20 years after the mine was closed and that the strata still are not stable even though the mining in the immediate vicinity was done 50 years ago, as is the case for the green sub-area. Table 1 indicates that the average annual rate of uplift for these four sub-areas varies between 6.2 mm/year (red sub-area) and 9.9 mm/year (blue sub-area). As mentioned earlier, the sub-areas were selected due to their relatively high uplift values. The maximum uplift is about 12 mm/year for the black and blue sub-areas. For the entire area presented in Figure 1b, the largest increase in uplift was 246.3 mm over the entire observation period or an average of 35.6 mm/year (coordinates of 11.210 km east-west and 6.757 km north-south).

Figure 5a presents the minimum, average, and maximum uplift for the blue sub-area, which is characterized by the largest uplift of all of the sub-areas that were considered. Figure 5b shows the curves for the brown sub-area, which corresponds to the lowest values of the uplift. Again, the variation over the entire observation period is presented without any extrapolation, and the origin of the time scale corresponds to the last panel mined in the vicinity. Figure 5b shows that the change in the movement of the surface over the first 19 months was negligible, i.e., between the start of the observation in December 2003 through July 2005. It was only after July 2005 that the average curve started to increase significantly. This also is visible in Figure 5a to a lesser extent, but the period is limited to just the first 12 months. This is in accordance with earlier research [13]. As stated above, the limited surface movement was observed during the period of 2000–2004, but obvious uplift has been occurring since 2005.

The variation between the minimum and maximum uplift as a function of time, or the so-called range, is relatively low, and it has been similar over the entire time period. For example, for the blue sub-area (Figure 5a), the range in uplift values has varied between 12.1 mm and 35.3 mm as a function of time, with an average range of 20.8 mm. On average, the maximum uplift has been 9.1 mm larger than the average, and the minimum uplift has been 11.7 mm smaller than the average. Over the entire time of the observation, the average rate of uplift was 9.9 mm/year, with the extremes being 7.7 and 12.0 mm/year. As discussed for the residual subsidence, the differences between the minimum and maximum uplifts are the result of both the inaccuracy of the measurement method and the impact of the spatial variation.

4. Discussion

The examples presented earlier clearly show that the movement of the surface does not stop a few years after coal seams are mined at relatively large depths using the longwall method with goaf. Without a doubt, this occurrence is not limited to the Belgian Campine coal basin; rather, it very likely is valid for all coal basins with similar conditions. The most relevant characteristics are (1) depths of −500 m and more, (2) the fact that several seams are mined above each other over depth intervals of several hundred meters, (3) that no chain pillars are used around the panels, and (4) that if a narrow zone of coal that is not mined between neighboring panels, it is crushed by the stress redistribution. So, for the case in which a single seam is mined at shallow depth with remaining chain pillars around the panels, it could be that the equilibrium already has been reached after a few years, but such conditions are not part of this research.

There probably are various reasons why it is almost always assumed that subsidence is limited in time. However, as explained above, this assumption is incorrect, at least for the conditions during the longwall mining in the Campine basin and in several other European coal basins with similar conditions. Probably, the most important reason is that, during the lifetime of a mine, it is very difficult, and may even be impossible, to distinguish the surface movements induced by the various longwall panels. At the end of mining a longwall panel, and sometimes even before the end, the mining of another panel in the vicinity is initiated, which also induces subsidence. This means that a point at the surface at a certain moment undergoes the impact of various longwall panels, i.e., those mined in the past (i.e., residual subsidence) and the one(s) currently being mined (initial and principal subsidence). Another reason is that, mainly in the past, people were less interested in the long-term impact of mining and gave little thought to the potential effects associated with the closure of a mine or the closure of a section of a mine. However, by introducing the concepts of sustainable mining, the long-term impact of mining on its surroundings has been receiving greater attention. This means that the period after the closure of a mine is a period that should not be neglected. It is obvious that during mining the amount of subsidence is largest. However, the impact of the smaller long-term surface movement rates on the already damaged or weakened buildings and infrastructure should not be neglected. In other words, a complete picture of the different phases is useful and necessary.

A reason which is more linked to practical considerations is that once the difference in surface movement between successive levelling campaigns becomes equal to the accuracy of the levelling method, or, in other words, the differences are very small (e.g., less than 10 or 20 mm), one has the tendency to stop monitoring altogether. In one way, this is understandable since such levelling campaigns are costly and time-consuming. Nevertheless, a significant advantage of satellite images is that these data are collected, independent of the size of movement or a specific need for monitoring. Often, one also assumes that low subsidence rates have no impact on structures. However, the impact of a subsidence of −0.5 m during a month, for example, probably is comparable to a subsidence rate of −50 mm/year for 10 years or of −5 mm/year for a century.

There also are some more general reasons why subsidence over very long periods has not been studied. Although research is being conducted on the time-dependent behavior of rock masses (e.g., Reference [18]), the time effect often is not included in an analysis, and hence, it is not well understood. For some types of rocks, e.g., for salt, clay, and shale, more research has been conducted on the time dependency because these rocks are sensitive to creep. However, the time intervals considered for these rocks are mostly in the range of days to months and are limited to the period immediately following the excavation. For example, Hawlader et al. [19] studied periods of up to 250 days for shaly rocks, and Bérest et al. [20] studied the impact of salt samples over nearly 700 days, but such studies are rather rare. Again, it is understandable that research projects that cover multiple years, or several decades, are not easy to organize, and the results only become available in the long-term. An interesting application where time periods of several centuries were studied was the failure of the historic masonry towers [21,22]. The collapse of these towers has been attributed to gradually increasing mechanical damage, such as creep cracking, which occurs under constant stress levels.

If masonry, with compressive strength of the same magnitude as weak-to-middle strong rocks and loaded by stress levels smaller than the strength, fails after several hundreds of years, one must assume that an assembly of broken and fractured weak rock blocks in the goaf zone will weaken further over a very long period. The flooding of the area also has an overall weakening effect on rock material, and two main mechanisms are involved. First, there is a direct weakening of various types of rock, e.g., shale, when they are submerged in water [23,24]. Wong et al. [23] concluded that, due to the considerable variations of texture and lithology, the extent of the water-weakening effect varies extensively among the different types of rocks, spanning from nearly negligible in quartzite to a 90% reduction of the uniaxial compressive strength in shale. They also specified for sedimentary rocks that, while sandstone mostly has losses in Young's modulus less than 20%, other sedimentary rocks have much higher losses. These losses in shale exhibit a normal distribution with a mean of around 50%. Some of the losses in the Young's modulus of mudstone can be 90% or more, while the losses of coal range between 30 and 40%. Second, an increase in pore pressure brings the rock closer to the failure envelope through the concept of effective stresses. In other words, additional fractures can be induced by the decrease of the effective stresses. Due to all of these mechanisms, one should not be surprised that residual subsidence is more prevalent than generally assumed.

5. Conclusions

The remote sensing data show, without a doubt, that the strata above longwall coal mining for the conditions of the Belgian Campine basin are not stabilized after a few months or years. Surface movements of a magnitude of about 10 mm/year have been measured several decades after mining [25].

There is a clear difference between the movement of the surface above the undermined area and the zones around this area.

This study and others also have shown that a new phenomenon in surface movement occurs after the closure of the mines and the flooding of the underground infrastructure and surrounding rock mass. It is important to note that the uplift is not a simple rebound of the past subsidence. The spatial variation of the residual subsidence is different from that of the uplift values [17]. Other studies of European coal basins have shown clearly that uplift is linked directly to flooding of the underground [26,27]. Currently, we conduct further studies of the mechanisms of this uplift, but it is most likely a combination of the impact of the water pressure on the effective stresses, and hence, on the relaxation of the rock, and it possibly could be due to the impact of the swelling of clay minerals. A question which cannot be answered at this time is whether the uplift is limited in time. In the eastern part of the Belgian Campine basin, uplift already has been occurring for about 30 years, and there are no signs that it will stop soon. However, if the phenomenon is governed by re-establishing the original distribution of the pore pressure, it can be concluded that the uplift will stop once this is done.

Another unanswered question is whether the measured uplift is the result of downward and upward movement, i.e., whether the residual subsidence is still ongoing. In fact, it can be assumed that there is no reason or indication why the residual subsidence would have stopped. This would mean that the real upward component in the resulting movement is about twice as large as the uplift values presented above. Then, the difference in the spatial distributions of the residual subsidence and uplift also would be larger.

Therefore, if the uplift component were to stop at a certain time, there is, in fact, no reason why the resulting movement would not be downward again, since there are no signs that the residual subsidence will stop soon. When the blocks of rock are submerged in water, they are weakened, and hence, the compression of the goaf material even can increase. In a similar way, the increase of the pore pressure brings the rock closer to the failure envelopes, resulting in a larger probability of fracturing, and thus, weakening.

Since we are confronted with a relatively new phenomenon, i.e., the occurrence of uplift above closed coal mines, it is logical that there are no complete answers to all of the questions regarding these observations at this stage. Thus, it is obvious that additional research is needed. So, if one would ask the question of how long the impact of coal mining on surface movement will last, one cannot answer it at this moment, but perhaps it could last even centuries. However, it is reasonable to conclude that it will be a period of time much longer than those that normally are used in the planning and design of rock mechanical excavations.

Mining always has a certain impact on the land use. The use of a total extraction method without any backfill always leads to a significant amount of surface movement. For shallow mining, this can mean in certain cases that the surface area can never be used again for building purposes, even sometimes for agricultural use. For deep mining, the consequences are not always so stringent. In the case of the Campine coal basin, mining took place under villages and towns. This created damage to buildings and infrastructure during the mining phase. However, as the surface movement does not stop after a few years, buildings are further deformed and, hence, the risk of additional damage remains. The large values of initial subsidence also have resulted in whole surface areas that would have been flooded, if the surface water had not been pumped out. And these areas can only remain dry by pumping out the surface water for ever. If one would stop these pumps, these areas would fast submerge. Additional surface movement over long time periods, either residual subsidence or uplift, will not lead to new types of restrictions to the land use, but the surface areas under threat of being submerged can increase or can be shifted, and the natural flow in small waterways can be affected too. So, even small rates of surface movements can have after several decades a significant impact on the required water management.

The main overall conclusion of the results presented in this paper is that both mining companies and legislators should be well aware of the long-term impact of mining and this should be taken into account when one defines the legal and the financial responsibilities of mining companies after closure.

Funding: This research received no external funding.

Conflicts of Interest: The author declares no conflict of interest.

References

1. Peng, S.S. *Surface Subsidence Engineering*; SME: New York, NY, USA, 1992.
2. Galvin, J.M. *Ground Engineering—Principles and Practices for Underground Coal Mining*; Springer International Publishing: Cham, Switzerland, 2016.
3. Huang, Y.; Zhang, J.; Yin, W.; Sun, Q. Analysis of overlying strata movement and behaviors in caving and solid backfilling mixed coal mining. *Energies* **2017**, *10*, 1057. [CrossRef]
4. Zhang, B.; Ye, J.; Zhang, Z.; Xu, L.; Xu, N. A comprehensive method for subsidence prediction on two-seam longwall mining. *Energies* **2019**, *12*, 3139. [CrossRef]
5. Vervoort, A. Various phases in surface movements linked to deep coal longwall mining: From start-up till the period after closure. *Int. J. Coal Sci. Technol.* **2020**. [CrossRef]
6. Luo, Y.; Peng, S.S. Long-term subsidence associated with longwall mining—Its causes, development and magnitude. *Min. Eng.* **2000**, *52*, 49–54.
7. Al Heib, M.; Nicolas, M.; Noirel, J.F.; Wojtkowiak, F. Residual subsidence analysis after the end of coalmine work. Example from Lorraine colliery, France. In Proceedings of the Symposium Post Mining, Nancy, France, 16–17 November 2005; pp. 1–9.
8. Cui, X.; Zhao, Y.; Wang, G.; Zhang, B.; Li, C. Calculation of residual surface subsidence above abandoned longwall coal mining. *Sustainability* **2020**, *12*, 1528. [CrossRef]
9. Kajzar, V. Geodetic and seismological observations applied for investigation of subsidence formation in the CSM mine (Czech Republic). *Min. Miner. Depos.* **2018**, *12*, 34–46. [CrossRef]
10. Van Tongeren, P.; Dreesen, R. Residual space volumes in abandoned coal mines of the Belgian Campine basin and possibilities for use. *Geol. Belg.* **2004**, *7*, 157–164.
11. Vervoort, A. Surface movement above an underground coal longwall mine after closure. *Nat. Hazards Earth Syst. Sci.* **2016**, *16*, 2107–2121. [CrossRef]
12. Pöttgens, J.J. Uplift as a result of rising mine waters. In *The Development Science and Art of Minerals Surveying, Proceedings of the 6th International Congress of the International Society for Mine Surveying, Harrogate, UK, 9–13 September 1985*; August Aimé Balkema: Rotterdam, The Netherlands, 1985; Volume 2, pp. 928–938.
13. Vervoort, A.; Declercq, P.Y. Surface movement above old coal longwalls after mine closure. *Int. J. Min. Sci. Technol.* **2017**, *27*, 481–490. [CrossRef]
14. Devleeschouwer, X.; Declercq, P.Y.; Flamion, B.; Brixko, J.; Timmermans, A.; Vanneste, J. Uplift revealed by radar interferometry around Liège (Belgium): A relation with rising mining groundwater. In Proceedings of the Symposium Post Mining, Nancy, France, 6–8 February 2008; pp. 1–13.
15. Marinkovic, P.; Ketelaar, G.; van Leijen, F.; Hanssen, R.F. INSAR quality control: Analysis of five years of corner reflector time series. In Proceedings of the Fringe 2007 Workshop, Frascati, Italy, 26–30 November 2007; pp. 1–8.
16. Agioutantis, Z.; Robertson, J.; Gollaher, G.; Schaefer, N.; Silva, J.; Tripolitsiotis, A.; Partsinevelos, P. Permanent scatterer radar interferometry as an effective structure deformation monitoring tool over undermined areas. In Proceedings of the 7th International Conference on Remote Sensing and Geoinformation of the Environment, Paphos, Cyprus, 18–21 March 2019; pp. 1–16. [CrossRef]
17. Vervoort, A. Long-term impact of coal mining on surface movement: Residual subsidence versus uplift. *Min. Rep. Glückauf* **2020**, *156*, 136–141.
18. Malan, D.F. Simulating the time-dependent behaviour of excavations in hard rock. *Rock Mech. Rock Eng.* **2002**, *35*, 225–254. [CrossRef]
19. Hawlader, B.C.; Lee, Y.N.; Lo, K.Y. Three-dimensional stress effects on time-dependent swelling behaviour of shaly rocks. *Can. Geotech. J.* **2003**, *40*, 501–511. [CrossRef]

20. Bérest, P.; Béraud, J.F.; Brouard, B.; Blum, P.A.; Charpentier, J.P.; de Greef, V.; Gharbi, H.; Valès, F. Very slow creep tests on salt samples. *Epj Web Conf.* **2010**, *6*, 22002. [CrossRef]
21. Verstrynge, E.; Schueremans, L.; Van Gemert, D.; Hendriks, M.A.N. Modelling and analysis of time-dependent behaviour of historical masonry under high stress levels. *Eng. Struct.* **2011**, *33*, 210–217. [CrossRef]
22. Verstrynge, E.; Van Gemert, D. Creep failure of two historical masonry towers: Analysis from material to structure. *Int. J. Mason. Res. Innov.* **2018**, *3*, 50–71. [CrossRef]
23. Wong, L.N.Y.; Maruvanchery, V.; Liu, G. Water effects on rock strength and stiffness degradation. *Acta Geotech.* **2016**, *11*, 713–737. [CrossRef]
24. Cai, X.; Zhou, Z.; Liu, K.; Du, X.; Zang, H. Water-weakening effects on the mechanical behavior of different rock types: Phenomena and mechanisms. *Appl. Sci.* **2019**, *9*, 4450. [CrossRef]
25. Vervoort, A.; Declercq, P.Y. Upward surface movement above deep coal mines after closure and flooding of underground workings. *Int. J. Min. Sci. Technol.* **2018**, *28*, 53–59. [CrossRef]
26. Caro Cuenca, M.; Hooper, A.J.; Hanssen, R.F. Surface deformation induced by water influx in the abandoned coal mines in Limburg, The Netherlands observed by satellite radar interferometry. *J. Appl. Geophys.* **2013**, *88*, 1–11. [CrossRef]
27. Samsonov, S.; d'Oreye, N.; Smets, B. Ground deformation associated with post-mining activity at the French—German border revealed by novel InSAR time series method. *Int. J. Appl. Earth Obs. Geoinf.* **2013**, *23*, 142–154. [CrossRef]

Article

Study on Foamed Concrete Used as Gas Isolation Material in the Coal Mine Goaf

Changyu Xu [1,2], Lijun Han [1,2,*], Maolin Tian [1,2], Yajie Wang [1] and Yuhao Jin [1,2]

[1] School of Mechanics and Civil Engineering, China University of Mining and Technology, Xuzhou 221116, China; TS18030059A31@cumt.edu.cn (C.X.); tianml@cumt.edu.cn (M.T.); TS17030122P3@cumt.edu.cn (Y.W.); jinyuhao@cumt.edu.cn (Y.J.)

[2] State Key Laboratory for Geomechanics and Deep Underground Engineering, China University of Mining and Technology, Xuzhou 221116, China

* Correspondence: hanlj@cumt.edu.cn

Received: 25 July 2020; Accepted: 22 August 2020; Published: 25 August 2020

Abstract: In view of the serious threat of gas accumulation in the coal mine goaf and the limitations of the existing gas sealing materials, the orthogonal experiment was developed to study a new type of foamed concrete for mine gas sealing. Dry density, gas permeability, and compressive strength were studied as the material indicators according to the demands of the gas isolation material in the coal mine goaf, and the experimental results showed that foam content was the most important factor. Meanwhile, the optimum mix was selected according to the influence of foam content as well as the engineering requirement. Then two application modes of this foamed concrete for goaf gas isolation were put forward, after which the convection-diffusion model of gas was built by COMSOL Multiphysics (COMSOL, Inc., Stockholm, Sweden) to reveal the mechanism of different application modes using the parameters of the new foamed concrete. Simulation results showed that this foamed concrete used as isolating material for goaf gas could significantly decrease the gas concentration in workface, which can provide a reference for similar engineering.

Keywords: foamed concrete; orthogonal experiment; optimum mix; coal mine goaf; gas isolation

---

1. Introduction

In recent years, China has widely used the comprehensive mechanized coal mining method, which has the characteristics of fast advancement speed and long workface. However, this method has a high degree of disturbance to the coal and rock mass, which can make the gas rapidly accumulate in the goaf and pose a serious threat to the safe production in a gas-rich coal mine [1–3]. Filling and sealing, as one of the key technologies to control the goaf gas, is a necessity to isolate the gassy areas and achieve permanent barriers between the workface and gas sources [4]. However, the existing inorganic gas-sealing materials are not efficient enough to seal the leaking gas, and the existing organic foam materials have negative impacts on the underground environment. Therefore, it is necessary to develop a low-cost, effective, and environmentally friendly material for roadway filling and sealing, which has the properties of compression, expansion, filling, sealing, and gastight, so as to ensure the safety, reliability, simplicity, and rapidity of coal mine underground work. Due to its low density, low thermal conductivity, low permeability, high expansibility, and high strength, foamed concrete has interested many experts [5,6]. Wang et al. [7] studied the effects of foam content, crumb rubber content, fly ash content, and water/binder ratio on the physical, mechanical, and waterproof properties of a lightweight foamed concrete by the single-factor and orthogonal tests. Ben et al. [8] investigated the mechanical characterization of non-autoclaved foam concrete according to its macro porosity. Khan et al. [9] investigated the effects of cement and recycled glass powder contents, water to cement ratio, and volume of foam on the plastic density, dry density, and compressive strength of foam concrete. Bing et al. [10] developed structural foamed concretes by using silica fume, fly ash, and polypropylene fiber and presented the use of fly ash for fully replacing sand to produce foamed concrete. Huang et al. [11] investigated the proportioning and properties of Portland cement-based ultra-lightweight foam concrete which was recommended to achieve energy efficiency in buildings. Zhao et al. [12] developed a new type of foamed concrete used as a seismic isolation material by orthogonal experiment. Hu et al. [13] developed a new type of foamed concrete and low-alkalinity sulphoaluminate cement, to control air leakage. However, the foamed concrete from the above studies is rarely applied in the coal mine goaf to control the gas. Some studies have studied the gas permeability of goaf or concrete: Dziurzynski et al. [14] developed a mathematical model for sealing mine goal based on the balance of the volume of a mixture supplied and contained in the body formed in the goaf. Based on the standard specimen test, Zhang et al. [15] studied the relationship between the gas permeability of slag high performance concrete and the carbonation depth, and the law of influence on carbonation depth of slag high performance concrete by the gas permeability is concluded. Md et al. [16] adopted two different procedures to carry out permeability tests on concrete specimens either unloaded or preloaded during the heating process, which showed that concrete permeability strongly depends on crack width and orientation.

Hence, this paper aims to determine the effect of mix proportions on the properties of foamed concrete and then produce a new type of foamed concrete that could be used as gas isolation material in the coal mine goaf. Firstly, the orthogonal experiment was developed to study the optimum mix, in which the dry density, gas permeability, and compressive strength were studied as material indicators. Then two application modes of the foamed concrete for gas isolation in a mining area were investigated. At last, taking a large-space goaf with high-strength and rapid advance as the engineering background, the convection-diffusion model of gas was built to reveal the mechanism of different application modes using the parameters of the new foamed concrete.

2. Material and Methods

The foamed concrete in this paper mainly consists of the foam, cement, water, and other admixtures. Foamed concrete has unique properties such as low density, low strength, high impact resistance, and thermal insulation. However, its expansibility and gas permeability are dominant when used as a gas isolation material in the coal mine goaf. The objective of the orthogonal experiment was thus to produce a new type of foamed concrete that had low density, certain strength, and low gas permeability.

2.1. Material

The materials used in this investigation mainly include the Portland cement (grade 42.5, China United Cement Group Co., Ltd., Xuzhou, China), naphthalene series superplasticizer, sodium silicate solution and concentrated high-efficiency cement foaming agent (Zhengzhou Pengyi Chemical Building Materials Co., Ltd., Zhengzhou, China). The properties of the Portland cement and the sodium silicate solution are shown in Tables 1 and 2.

Table 1. Indexes of ordinary Portland cement.

Fineness/%	Initial Setting Time/min	Final Setting Time/min	Compressive Strength/MPa		
			3 days	14 days	28 days
2.4	162	247	4.6	36.4	45.6

Table 2. Parameters of sodium silicate solution.

Baume Degree/°Be	Modulus	Density/(g/mL)	SiO_2/%	Na_2O/%	Fe/%
40	3.2	1.37	26.3	8.6	0.5

2.2. Design of the Orthogonal Experiment

The four main factors of foamed concrete that can affect its properties were chosen according to the experimental goal: water cement ratio, foam (by volume), water reducing agent (by weight), and sodium silicate solution (by weight) based on 1 kg mass of solid powder. For each component, there were four levels, and thus, in total, 16 concrete series were investigated. According to the early analysis of the single factor influence of foamed concrete, the most suitable proportioning interval is determined, as shown in Table 3.

Table 3. Parameters for orthogonal experiment.

Level	Water Cement Ratio	Foam/L	Water Reducing Agent/%	Sodium Silicate Solution/%
1	0.40	0.5	0.2	0.5
2	0.45	1.0	0.4	1.0
3	0.50	1.5	0.6	2.0
4	0.55	2.0	0.8	3.0

2.3. Specimen Preparation

As shown in Figure 1, the specimen preparation can be summarized as follows: firstly, the foaming agent was diluted with water at a ratio of 1:20 (by volume) and then added to the foaming machine. Secondly, the Portland cement, water-reducing agent, and sodium silicate solution are mixed evenly according to a certain mix proportion to make a cement slurry of certain water cement ratio, and then the foam produced in the first step was added and mixed until a uniform slurry was produced. Thirdly, the specimens were poured into molds and compacted with an external vibrator, demolded after 24 h, and kept in a box at constant temperature (22 ± 2 °C) and humidity (95%) up to the day of testing.

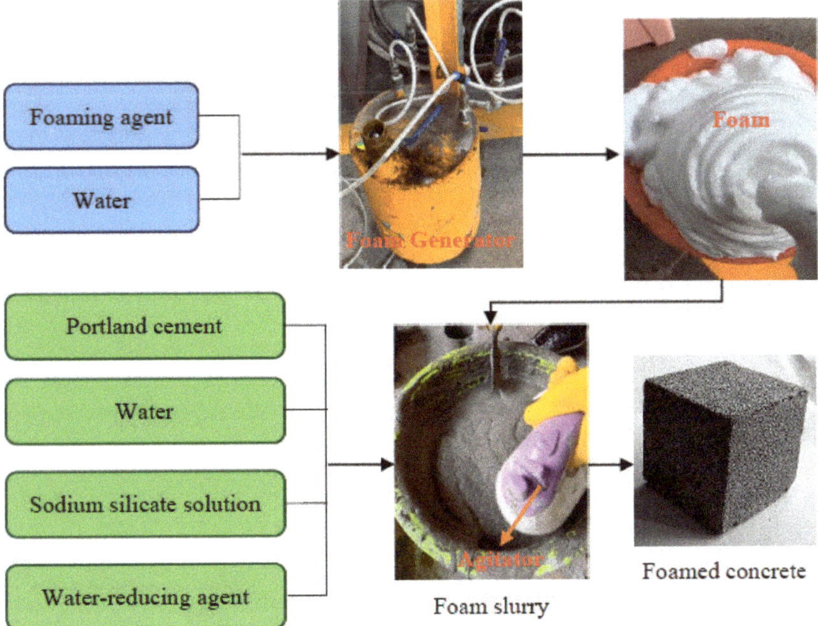

Figure 1. Procedure of specimen perpetration.

2.4. Test Contents and Methods

2.4.1. Dry Density

In order to save the engineering cost, foamed concrete needs to be of good expansibility, and dry density was chosen as the evaluation index of the expansibility, which is convenient for measurement. The standard cylindrical samples of 50 mm × 100 mm were selected, put into a constant temperature vacuum drying oven (see Figure 2), adjusted to 80 ± 5 °C for 24 h, and then the temperature adjusted to 120 °C until gaining the constant weight M. The diameter and height of the sample were measured, and its volume was calculated V. Then the dry density ρ of the concrete material can be obtained by the formula:

$$\rho = \frac{M}{V} \times 1000 \tag{1}$$

where, ρ is the dry density of the sample, kg/m^3; M is the mass of the sample after drying, g; V is the volume of the sample, cm^3.

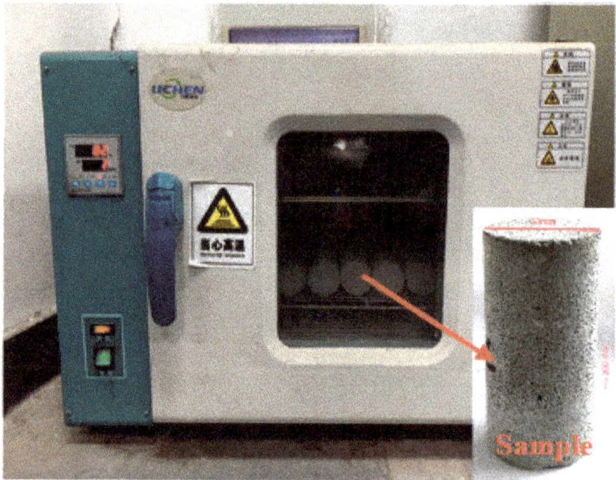

Figure 2. Constant temperature vacuum drying oven.

2.4.2. Gas Permeability

The foamed concrete is mainly used to isolate gas in the coal mine goaf, so the gas permeability is used as anther evaluation index, which can be tested by gas permeability measurement system (see Figure 3). First, 50 mm of the samples was cut off in dry density test, a layer of epoxy resin was applied on the outside of samples, and the sample was put into the concrete gas penetration tester, ensuring the airtightness of the device. Then the air pressure was controlled to be about 0.1–0.2 MPa at the inlet in the tester, so that the gas entered into the sample from the lower surface and exited from the upper surface. The air inlet pressure P_1 and air outlet pressure P_2 were recorded under different air supply pressures. Then, the gas permeability of the sample can be easily obtained from the apparent gas permeability k_a, which can be calculated as follows:

$$k_a = \frac{Q}{A} \frac{2\eta h P_1}{P_1^2 - P_2^2} \tag{2}$$

where, k_a is the apparent gas permeability, m^2; Q is the gas flow, mL/min; A is the cross section area of the sample, mm^2; η is the viscosity of the compressed gas, Pa·s; h is the height of the test piece, mm; P_1 is the inlet pressure, bar; P_2 is the outlet pressure, bar.

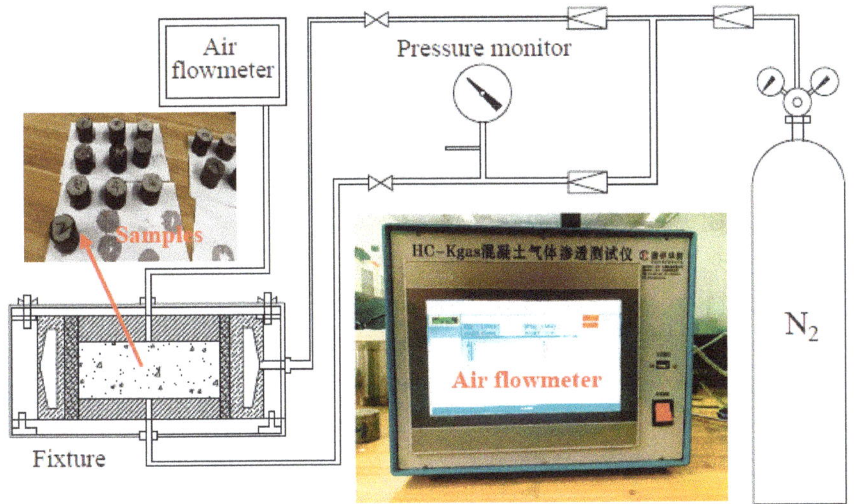

Figure 3. Gas permeability measurement system.

2.4.3. Compressive Strength

Although there is less requirement for the compressive strength of the foamed concrete, the higher the compressive strength, the more stable the whole goat can be maintained, and the more conducive to the mining of the adjacent workface. Considering the short time filling effect and long time supporting effect, the compressive strength of 3 days and 28 days of the foamed concrete were selected as the other evaluation indexes, which can be tested by universal testing machine CSS-14100 (see Figure 4). Cube samples of 70.7 × 70.7 × 70.7 mm were taken, their dimensions measured, and the contact area A_1 was calculated. Then the sample was placed on the press plate center of the universal testing machine. The testing machine was started and controlled with a displacement loading speed of 0.02 mm/s. When the specimen cracks and reaches the complete failure state, the test is finished and the data as well as the maximum axial pressure F is recorded in the computer. Then the compressive strength σ of the foamed concrete can be obtained by the formula:

$$\sigma = \frac{F}{A_1} \tag{3}$$

where, σ is the compressive strength of the sample, MPa; F is the maximum failure load of the sample, N; A_1 is the contact area of the sample, mm^2.

As a result, the dry density, gas permeability, compressive strength of 3 days and 28 days, which were used as four test indexes according to the demands of the foamed concrete for gas isolation, were obtained by the methods above [17,18].

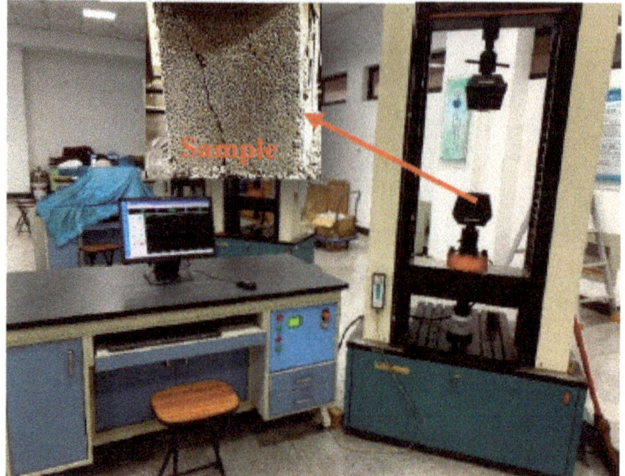

Figure 4. Universal testing machine.

3. Results and Optimum Mix

3.1. Test Results and Analysis

Average level is the average value of each factor at the same level. The average level of four indicators (dry density, gas permeability, compressive strength of 3 days and 28 days) under different factors can be seen in Figure 5a–d, and it can be noted that the foam is the main factor influencing the four indexes as mentioned above. The dry density and compressive strength drop and the gas permeability increases as the volume of foam increases. In order to further investigate the mechanism of this phenomenon, a microscope was used to magnify the samples with different foam content for 500 times, as shown in Figure 6.

When the foamed concrete sample is magnified by microscope, it can be seen that when the amount of foam is less, there are fewer pores in the sample, and these pores are small, airtight, and smooth. With the increase of foam content, both the number and the diameter of pores increase significantly. When the amount of foam reaches 2 L, the pores in the samples are denser and larger, and the connected pores are formed, which leads to larger gas permeability of the foamed concrete. At the same time, since the fact that foam helps to produce more pores, the increase of foam content can lead to a decrease in other solid materials in the unit volume, resulting in a decrease in density and strength.

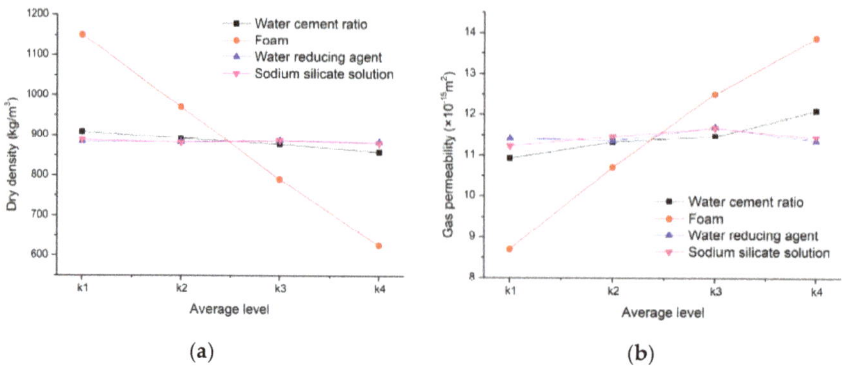

(a)

(b)

Figure 5. *Cont.*

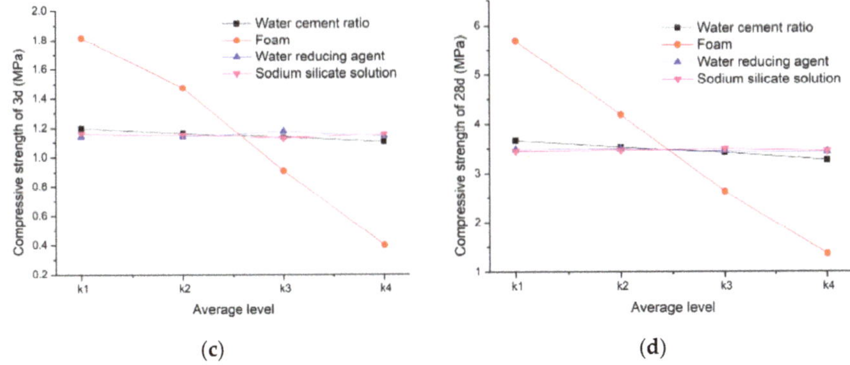

(c)   (d)

Figure 5. Average level of indicators under different factors: (**a**) average level of dry density; (**b**) average level of gas permeability; (**c**) average level of compressive strength of the 3rd day foamed concrete; (**d**) average level of compressive strength of the 28th day foamed concrete.

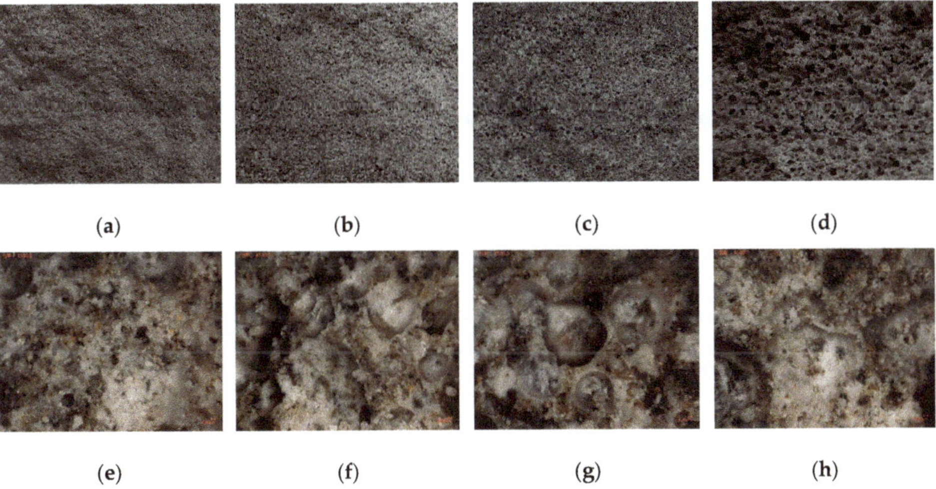

(a)   (b)   (c)   (d)

(e)   (f)   (g)   (h)

Figure 6. Pore structures of different foam content in 1 kg mass of solid powder: (**a**) 0.5 L foam; (**b**) 1.0 L foam; (**c**) 1.5 L foam; (**d**) 2.0 L foam; (**e**) 0.5 L foam (×500 times); (**f**) 1.0 L foam (×500 times); (**g**) 1.5 L foam (×500 times); (**h**) 2.0 L foam (×500 times).

3.2. Effect of Foam Content on Properties of Foamed Concrete

In order to further verify the effect of foam content on the properties of foamed concrete, the samples of 0.5, 1.0, 1.5, and 2.0 L foam based on 1 kg mass of solid powder with a water cement ratio of 0.40 was prepared, and the change rules of gas permeability, dry density, and compressive strength were studied. As shown in Figure 7, when other conditions remain unchanged, the dry density and compressive strength decrease with the increase of foam content, and the gas permeability increases with the increase of foam content, which is consistent with the above results.

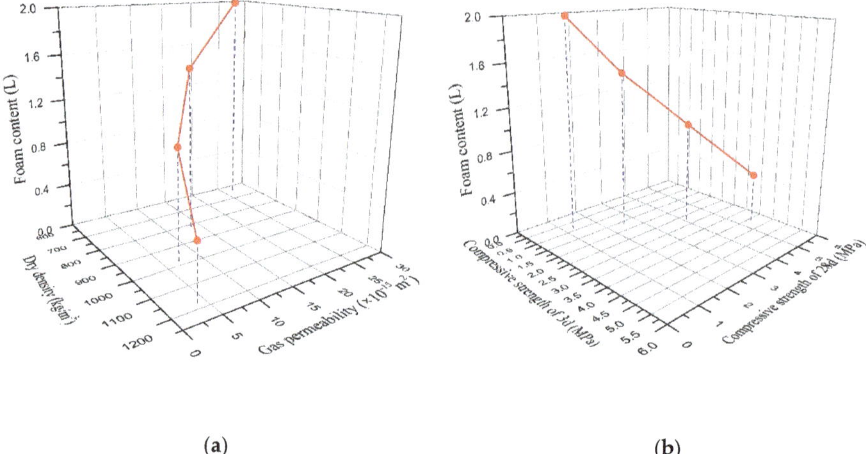

(a)   (b)

Figure 7. (**a**) The effect of foam content on dry density and gas permeability; (**b**) the effect of foam content on compressive strength of 3d and 28d.

3.3. Optimum Mix

The comprehensive balance method was used to compare the results of each factor. As the effect of foam content on foamed concrete is far greater than the other three factors, we first determined the foam content. Due to the large consumption of foamed concrete in the coal mine goaf, considering the economy, Level 1 and Level 2 were excluded according to the dry density in Figure 5a. Meanwhile, considering the stability of the foamed concrete after consolidation, it can be seen that the compressive strength of Level 4 is far below the average level according to Figure 5c,d, so the foam content was selected at Level 3.

As can be seen from Figure 5, the water cement ratio is the secondary factor. In order to isolate gas better, Level 4 with highest gas permeability was first eliminated according to Figure 5b. Considering the economy, the water cement ratio should be as large as possible, so Level 3 was selected according to Figure 5a.

The above analysis can preliminarily determine the selection of the 11th mix proportion of 16 concrete series. In the 11th mix proportion, water reducing agent is at Level 1 and sodium silicate solution is at Level 2, which is very economical. According to the research results and the actual engineering requirements, the 11th mix proportion was chosen to be the optimum mix which can be seen in Table 4, as well as its engineering properties.

Table 4. Proportions and properties of the new foamed concrete.

Optimum Mix Proportions		Engineering Properties	
Water cement ratio	0.5	Dry density/kg·m^{-3}	781
Foam/L	1.5	Gas permeability/×10^{-15} m^2	12.13
Water reducing agent/%	0.2	Compressive strength of 3 days/MPa	0.90
Sodium silicate solution/%	1.0	Compressive strength of 28 days/MPa	2.56

4. Application Modes of the Foamed Concrete

4.1. Application Modes

For the gas-rich coal mine, the gas flows into the goaf from the coal wall, the falling coal and the roof, which makes a large number of high concentration gas accumulated in the goaf, causing the gas leakage into the workface. Under normal conditions, the air pressure between the goaf and workface is basically balanced, and the gas in the goaf is mainly brought out from the upper corner by the roadway air from the lower corner, as shown in Figure 8. In order to reduce the gas concentration and ensure the safety and stability during the operation of the workface, two ways of isolating gas by using the foamed concrete are put forward:

- Sealing the gas, that is separating the goaf gas from the workface air by building the gas separation walls on both lower and upper corners.
- Replacing the gas, that is using the foamed concrete to replace the gas in the goaf void, after which the void can be filled and there is little space for gas accumulation.

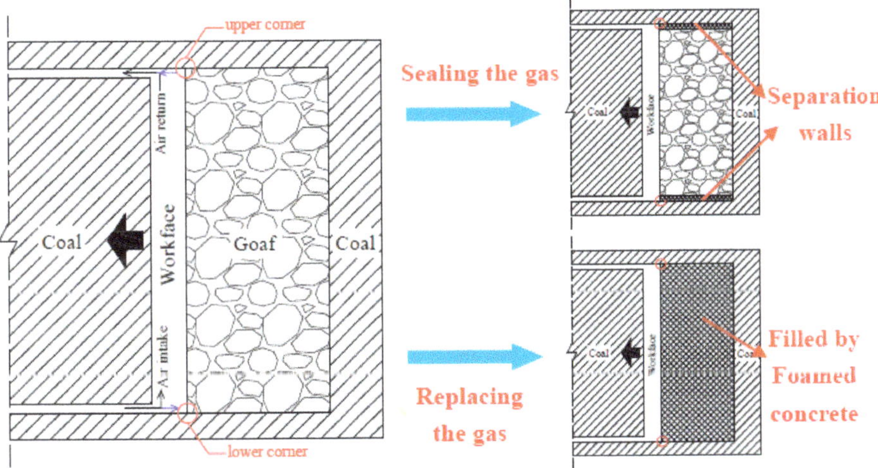

Figure 8. Schematic diagram of application modes.

4.2. Mechanism of Application Modes

4.2.1. Establish Mathematical Model

According to the characteristics of gas migration in the large-space goaf under the condition of high-intensity and rapid mining advance, the Brinkman seepage equation and Fick diffusion equation can be coupled to solve the transient process of gas diffusion under the action of air flow, so as to study the mechanism of the two application modes.

The convection and migration process of gas diffusion in goaf includes two physical processes: gas ventilation convection and concentration diffusion [19,20]. The gas diffusion equation conforms to Fick diffusion law, that is

$$\delta t \frac{\partial C}{\partial t} + \nabla \cdot (-D \nabla C) + v \cdot \nabla C = R \tag{4}$$

where, δt is the instantaneous time proportion coefficient, which is equivalent to porosity; C is the gas concentration, mol/m^3; ∇ is the lalpace operator; D is the diffusion coefficient, m^2/s; v is the average velocity, m/s; R is the source term, mol/(m^3·s).

Under the condition of fully mechanized coal mining, the goaf caving area is composed of broken caving rock mass, which can be regard as the large porous medium. Its compaction degree is related to the support pressure above the goaf, in which the air flow channel system is relatively complex, and the Brinkman seepage equation is more suitable, that is

$$\frac{\eta}{k} \cdot v = \nabla \cdot (-pI + \eta(\nabla v + (\nabla v)^T)) + F \tag{5}$$

where, v is the flow velocity of fluid, m/s; k is the permeability, m^2; η is the dynamic viscosity coefficient, Pa·s; p is the fluid pressure, Pa; I is the unit vector; F is the fluid resistance.

In the calculation model of this paper, the simultaneous equations (Equations (4) and (5)) can solve the transient process of gas diffusion under the action of air flow, which can be calculated by numerical simulation software COMSOL Multiphysics.

COMSOL Multiphysics is a large-scale advanced numerical simulation software developed by the Swedish company COMSOL. Based on the finite element theory, COMSOL Multiphysics is developed to simulate various physical phenomena, which can better simulate the coupling of multiple physical fields and describe physical phenomena by solving various mathematical models [19].

4.2.2. Model Calculation

Referring to the actual size of the comprehensive mechanized coal mining workface, the calculation model of 80 × 195 m was established by COMSOL Multiphysics (in order to simplify the calculation, this paper only studied the goaf area formed by about one week of coal mining without considering the dynamic process). As shown in Figure 9, the width of the roadway is 5.6 m, so the lower 5.6 m part of the left side is treated as the boundary of air inlet, and the upper 5.6 m part of the left side is treated as the boundary of air outlet, and the other boundaries are airtight. Since the air flow in the roadway of workface accounts for the majority, and the air flow penetrating into the goaf only accounts for the minority, the pressure difference between air inlet and outlet is set as 100 Pa. At the same time, the air inlet and outlet are set as the convection boundary, between which is the airtight boundary, and the gas concentration flux was supplied by the other boundaries. There are 1 atm and 3 mol/m^3 initial concentration gas in the goaf, and the other numerical simulation parameters are shown in Table 5.

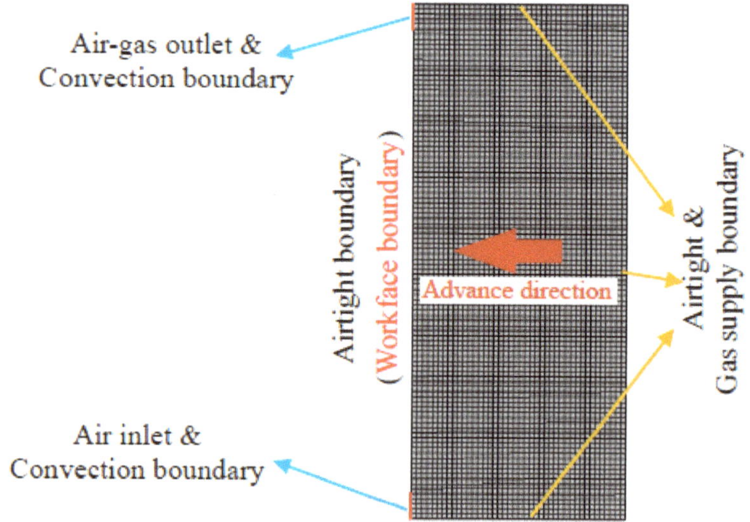

Figure 9. Computational mode and its boundary conditions.

Table 5. Numerical simulation parameters.

Parameters	δt	D (m^2/s)	R mol/(m^2·s)	η (Pa·s)	k (m^2)
Value	0.55	2×10^{-5}	3×10^{-6}	1.8×10^{-5}	3.24×10^{-8}

When simulating 'gas sealing', the gas separation walls with equal width of roadways were arranged on both sides of the goaf, the parameters of which were set as the foamed concrete in Table 4, and the remaining parts of the model were unchanged. When simulating 'gas replacement', the foamed concrete was used to block the entire cavity of the goaf and replace the gas in the empty hole, which could completely block the gas source. Since the replacement process was dynamic, the parameters of the whole calculation range were constantly changing. In order to simplify the study, the pressure difference between the air inlet and outlet was increased by five times in consideration of the displacement effect of the gas when the filling material was pumped, after which the parameters of the whole goaf were set as the foamed concrete in Table 4 [21–23].

4.3. Analysis of Simulation Results

4.3.1. No Gas Isolation

We drew the gas concentration distribution map in the goaf when the time was 0, 0.5, 1, 3, and 5 days, as shown in Figure 10a–e. The results show that under the condition of 100 Pa air pressure difference, it will take a certain time to start dispersing the gas because of the existence of 1 atm and 3 mol/m^2 initial concentration in the goaf. With the development of time, from the air inlet to the outlet, a fan-shaped concentration reduction area was gradually formed until it became stable about 5 days later, forming an accumulation area in the upper corner and keeping stable at 5–6 mol/m^2. If measures are not taken to isolate gas, the gas accumulation area near the upper corner will make the gas continuously flow out with the air, causing the hidden danger to the workface.

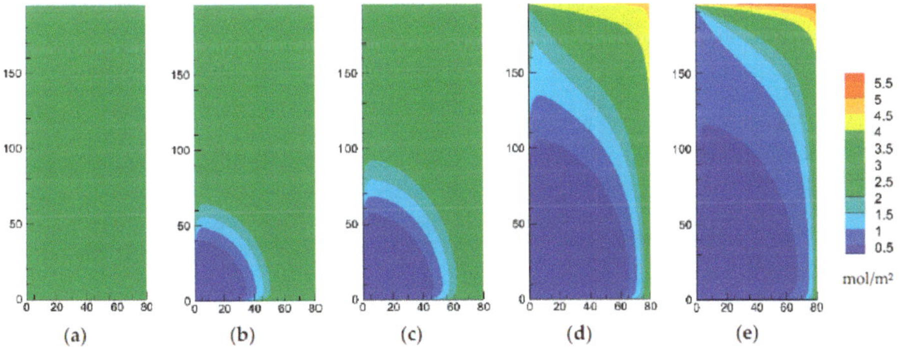

Figure 10. Concentration distribution in gas evacuation process: (**a**) $t = 0$ day; (**b**) $t = 0.5$ day; (**c**) $t = 1$ day; (**d**) $t = 3$ days; (**e**) $t = 5$ days.

4.3.2. Sealing the Gas

We drew the gas concentration distribution map in the goaf when the time was 0, 0.5, 2, 4, and 6 days, as shown in Figure 11a–e. Due to the existence of gas separation walls, the goaf is like a closed box, which is weakly affected by the air in roadways. Therefore, the gas accumulates continuously in the goaf instead of leaking into the workface. In addition to forming a low gas area near the air inlet, the gas concentration in most areas increases with time. It in dicates that the gas is sealed by the separation walls at two ends of the goaf instead of leaking into the workface, which ensures the safety and stability of the working face.

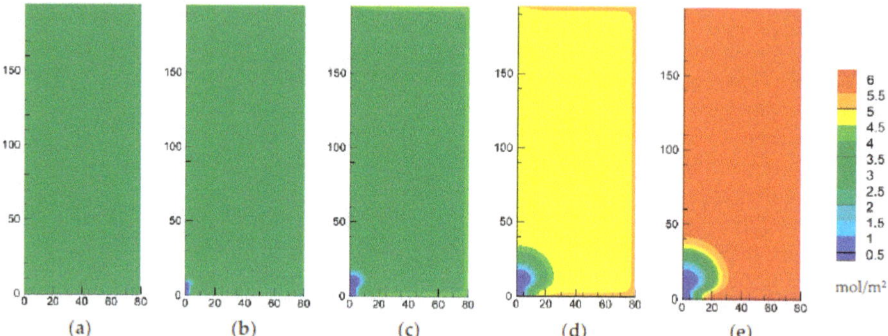

Figure 11. Concentration distribution in gas sealing process: (**a**) $t = 0$ day; (**b**) $t = 0.5$ day; (**c**) $t = 2$ days; (**d**) $t = 4$ days; (**e**) $t = 6$ days.

4.3.3. Replacing the Gas

We drew the gas concentration distribution map in the goaf when the time was 0, 0.125, 0.5, 1, and 3 days, as shown in Figure 12a–e. In the process of gas replacement by the foamed concrete, the gas is expelled into the air-return roadway from the outlet in a short time. Then the goaf become dense by filling with foamed concrete, so there is no new gas accumulation and the gas began to stabilize at a low level from the third day.

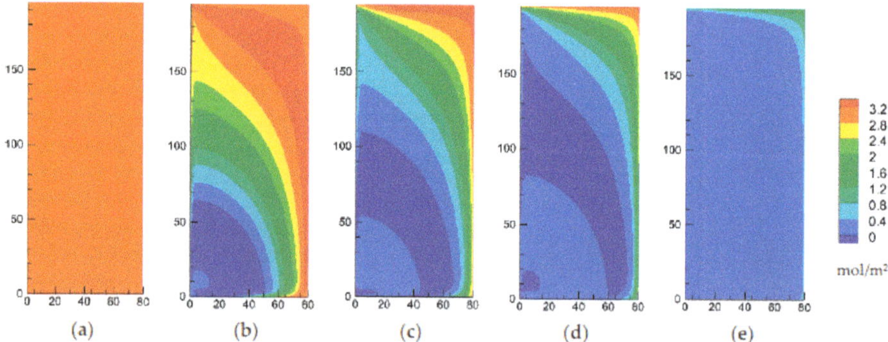

Figure 12. Concentration distribution in gas replacement process: (**a**) $t = 0$ day; (**b**) $t = 0.125$ day; (**c**) $t = 0.5$ day; (**d**) $t = 1$ day; (**e**) $t = 3$ days.

In this case, the gas in the workface will increase in a short time and the proper ventilation in roadways is necessary to carry out to remove the temporary high concentration gas near the outlet, after which the workface can be safely mined without any hidden danger of gas leakage.

4.3.4. Gas isolation Mechanism

Sealing and replacing are two very different methods of gas isolation. The former prevents the gas from leaking to workface by plugging the gas inlet and outlet using the foamed concrete. The latter firstly evacuates the existing gas, and then replaces the gas position with foamed concrete so as to fundamentally prevent gas leakage. The distribution curves of gas concentration along the workface boundary at different application modes is shown in Figure 13. The figure shows that when sealing the gas, since there is little influence of the air in roadway, only about 25 m range of the workface around the air inlet has the phenomenon of concentration reduction, and the gas concentration in the remaining part of workface exceeds the initial concentration and increases with time. However, when replacing the gas, the gas concentration of the workface boundary rapidly reduces to 0 near the inlet. The gas concentration along this boundary is lower than the initial concentration and decreases with the increase of time, until the entire boundary (except the outlet) remains at a very low level.

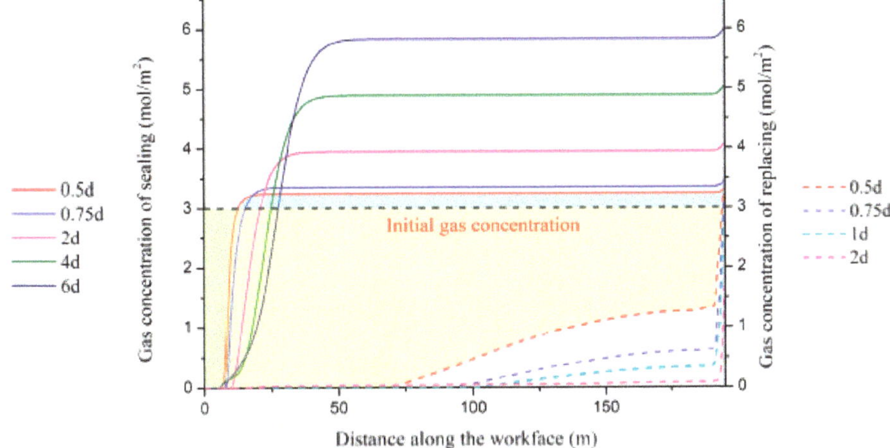

Figure 13. Concentration distribution along the workface.

5. Discussion and Conclusions

The mix proportions have a significant effect on the properties of foamed concrete, and the foam content is the most important factor. Based on the orthogonal experiment results, the optimum mix proportions were obtained, and a new type of foamed concrete was produced, which has low dry density, low gas permeability, and certain strength.

This new concrete material can be used to isolate the goaf gas in two forms: sealing the gas and replacing the gas. According to the convection-diffusion model of gas, the multi-field physic coupled simulation software COMSOL Multiphysics was utilized to conduct the numerical simulation. The simulation results indicated that when the foamed concrete is used to build air separation walls at both ends of goaf to seal the gas, the gas concentration in most of mined-out area increases with time except for a low gas area near the air inlet. When the foamed concrete is used to replace the gas in the goaf void, the gas can be expelled from the outlet in a short time, and then the gas concentration in the goaf is stable at a low level.

In view of its good gas isolation properties and the effective application modes, it is suitable and feasible to use this new type of foamed concrete as gas isolation material in the coal mine goaf, which can provide a reference for similar engineering.

Author Contributions: Conceptualization, L.H.; Formal analysis, C.X.; Investigation, M.T.; Methodology, Y.W.; Software, Y.J.; Writing—original draft, C.X.; Writing—review and editing, L.H. All authors have read and agreed to the published version of the manuscript.

Funding: This work was supported by the National Natural Science Foundation of China (grant number 51574223).

References

1. Zhang, J.; Cliff, D.; Xu, K.; You, G. Focusing on the patterns and characteristics of extraordinarily severe gas explosion accidents in chinese coal mines. *Process Saf. Environ. Prot.* **2018**, *117*, 390–398. [CrossRef]
2. Shi, S.; Jiang, B.; Meng, X.; Yang, L. Fuzzy fault tree analysis for gas explosion of coal mining and heading faces in underground coal mines. *Adv. Mech. Eng.* **2018**, *10*, 1687814018792318. [CrossRef]

3. Gao, Y.; Fu, G.; Nieto, A. A comparative study of gas explosion occurrences and causes in china and the united states. *Int. J. Min. Reclam. Environ.* **2015**, *30*, 269–278. [CrossRef]

4. Wang, C.; Liu, Y.; Hu, H.; Li, Y.; Lu, Y. Study on filling material ratio and filling effect: Taking coarse fly ash and coal gangue as the main filling component. *Adv. Civ. Eng.* **2019**, *2019*, 2898019. [CrossRef]

5. Amran, Y.H.M.; Farzadnia, N.; Abang Ali, A.A. Properties and applications of foamed concrete: A review. *Constr. Build. Mater.* **2015**, *101*, 990–1005. [CrossRef]

6. Ramamurthy, K.; Kunhanandan Nambiar, E.K.; Indu Siva Ranjani, G. A classification of studies on properties of foam concrete. *Cem. Concr. Compos.* **2009**, *31*, 388–396. [CrossRef]

7. Wang, R.; Gao, P.; Tian, M.; Dai, Y. Experimental study on mechanical and waterproof performance of lightweight foamed concrete mixed with crumb rubber. *Constr. Build. Mater.* **2019**, *209*, 655–664. [CrossRef]

8. Ben Youssef, M.; Miled, K.; Néji, J. Mechanical properties of non-autoclaved foam concrete: Analytical models vs. Experimental data. *Eur. J. Environ. Civ. Eng.* **2017**, *24*, 472–480. [CrossRef]

9. Khan, Q.S.; Sheikh, M.N.; McCarthy, T.J.; Robati, M.; Allen, M. Experimental investigation on foam concrete without and with recycled glass powder: A sustainable solution for future construction. *Constr. Build. Mater.* **2019**, *201*, 369–379. [CrossRef]

10. Bing, C.; Zhen, W.; Ning, L. Experimental research on properties of high-strength foamed concrete. *J. Mater. Civ. Eng.* **2012**, *24*, 113–118. [CrossRef]

11. Huang, Z.; Zhang, T.; Wen, Z. Proportioning and characterization of portland cement-based ultra-lightweight foam concretes. *Constr. Build. Mater.* **2015**, *79*, 390–396. [CrossRef]

12. Zhao, W.S.; Chen, W.Z.; Tan, X.J.; Huang, S. Study on foamed concrete used as seismic isolation material for tunnels in rock. *Mater. Res. Innov.* **2013**, *17*, 465–472. [CrossRef]

13. Wen, H.; Fan, S.; Zhang, D.; Wang, W.; Guo, J.; Sun, Q. Experimental study and application of a novel foamed concrete to yield airtight walls in coal mines. *Adv. Mater. Sci. Eng.* **2018**, *2018*, 1–13. [CrossRef]

14. Dziurzynski, W.; Krach, A.; Palka, T. Computer simulation of the impact of fed-in mineral substances on air and methane flow in longwall goaf. *Arch. Min. Sci.* **2010**, *55*, 517–535.

15. Zhang, X.; Zhou, X.; Zhou, H.; Gao, K.; Wang, Z. Studies on forecasting of carbonation depth of slag high performance concrete considering gas permeability. *Appl. Clay Sci.* **2013**, *79*, 36–40. [CrossRef]

16. Md, J.; Hatem, K.; Hélène, C.; Pierre, P.; Christian, L. The effect of compressive loading on the residual gas permeability of concrete. *Constr. Build. Mater.* **2019**, *217*, 12–19.

17. Kearsleya, E.P.; Wainwright, P.J. The effect of porosity on the strength of foamed concrete. *Cem. Concr. Res.* **2001**, *32*, 233–239. [CrossRef]

18. Nambiar, E.K.K.; Ramamurthy, K. Air-void characterisation of foam concrete. *Cem. Concr. Res.* **2007**, *37*, 221–230. [CrossRef]

19. Yang, T.; Li, B.; Ye, Q. Numerical simulation research on dynamical variation of permeability of coal around roadway based on gas-solid coupling model for gassy coal. *Int. J. Min. Sci. Technol.* **2018**, *28*, 925–932. [CrossRef]

20. Yang, T.H.; Chen, S.K.; Zhu, W.C.; Huo, Z.G.; Jiang, W.Z. Numerical model of nonlinear flow-diffusion for gas mitigation in goaf and atmosphere. *J. China Coal Soc.* **2009**, *34*, 771–777.

21. Wang, Y.; Luo, G.; Geng, F.; Li, Y.; Li, Y. Numerical study on dust movement and dust distribution for hybrid ventilation system in a laneway of coal mine. *J. Loss Prev. Process Ind.* **2015**, *36*, 146–157. [CrossRef]

22. Che, Q. Study on Coupling Law of Mixed Gas Three-Dimensional Multi-Field in Goaf. Ph.D. Thesis, China University of Mining and Technology, Beijing, China, 2010.

23. Liu, M.X.; Shi, G.Q.; Guo, Z.; Wang, Y.M.; Ma, L.Y. 3-D simulation of gases transport under condition of inert gas injection into goaf. *Heat Mass Transf.* **2016**, *52*, 2723–2734. [CrossRef]

Article

Water Preservation and Conservation above Coal Mines Using an Innovative Approach: A Case Study

Yujun Xu [1], Liqiang Ma [1,2,*] and Yihe Yu [1]

[1] State Key Laboratory of Coal Resources and Mine Safety, School of Mines, China University of Mining and Technology, Xuzhou 221116, China; xyj@cumt.edu.cn (Y.X.); yyhcmsj@cumt.edu.cn (Y.Y.)
[2] School of Energy, Xi'an University of Science and Technology, Xi'an 710054, China
* Correspondence: ckma@cumt.edu.cn; Tel.: +86-136-4520-1296

Received: 9 May 2020; Accepted: 28 May 2020; Published: 2 June 2020

Abstract: To better protect the ecological environment during large scale underground coal mining operations in the northwest of China, the authors have proposed a water-conservation coal mining (WCCM) method. This case study demonstrated the successful application of WCCM in the Yu-Shen mining area. Firstly, by using the analytic hierarchy process (AHP), the influencing factors of WCCM were identified and the identification model with a multilevel structure was developed, to determine the weight of each influencing factor. Based on this, the five maps: overburden thickness contour, stratigraphic structure map, water-rich zoning map of aquifers, aquiclude thickness contour and coal seam thickness contour, were analyzed and determined. This formed the basis for studying WCCM in the mining area. Using the geological conditions of the Yu-Shen mining area, the features of caved zone, water conductive fractured zone (WCFZ) and protective zone were studied. The equations for calculating the height of the "three zones" were proposed. Considering the hydrogeological condition of Yu-Shen mining area, the criteria were put forward to evaluate the impact of coal mining on groundwater, which were then used to determine the distribution of different impact levels. Using strata control theory, the mechanism and applicability of WCCM methods, including height-restricted mining, (partial) backfill mining and narrow strip mining, together with the applicable zone of these methods, were analyzed and identified. Under the guidance of "two zoning" (zoning based on coal mining's impact level on groundwater and zoning based on applicability of WCCM methods), the WCCM practice was carried out in Yu-Shen mining area. The research findings will provide theoretical and practical instruction for the WCCM in the northwest mining area of China, which is important to reduce the impact of mining on surface and groundwater.

Keywords: water protection; water-conservation coal mining (WCCM); influencing factors; "five maps; three zones; and two zoning plans"; water conductive fractured zone (WCFZ)

1. Introduction

In China, coal resources are mainly distributed in the north and west, especially in the provinces of Shanxi, Shaanxi, Inner Mongolia, and Xinjiang, which account for about 70% of all Chinese coal reserves. As the major coal production region, five large coal production bases have been developed in Northwest China, i.e., Shaanbei, Huanglong, Shendong, Ningdong, and Xinjiang (Figure 1). These bases contribute to over one third of China's annual coal production, providing strong support to the country's economic growth [1].

Northwest China has a dry climate and sparse vegetation, hence water resources are scarce, accounting for only 3.9% of the country's total [2,3]. Shallow groundwater, a key water source for local vegetation [4,5], is likely to be affected by mining, making the fragile ecosystems even more vulnerable [6–13]. In the Yu-Shen mining area, in the major production area of the Shaanbei base, the groundwater table has declined by more than 15 m over an area of 306.8 km^2 and by 8–15 m in a further 352.1 km^2 area. High intensity mining is found to be the direct cause of significant groundwater table decline in 71.5% of these areas [11]. It is clear that there are severe conflicts between water scarcity, ecological vulnerability, and coal resources exploitation.

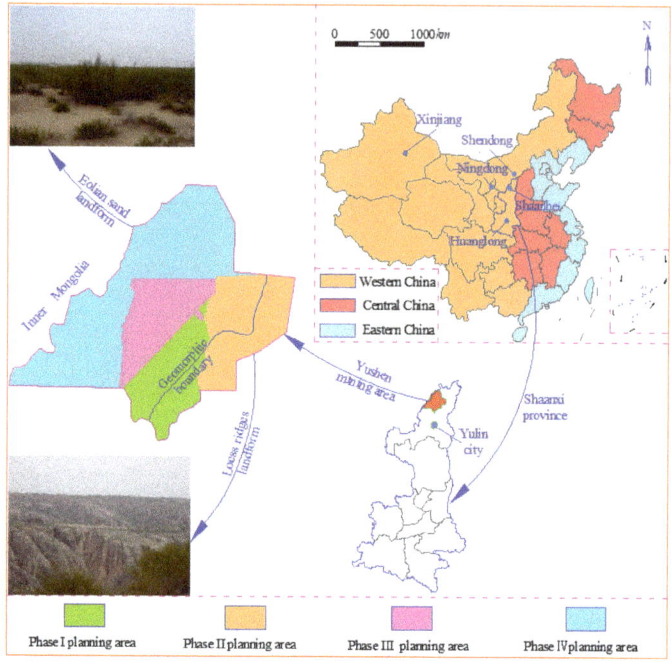

Figure 1. Location and landform of Yu-Shen mining area and five large coal production bases.

Researchers have proposed the concept of water-conservation coal mining (WCCM) and domestic and foreign researchers have conducted in-depth research on mining-induced groundwater loss and WCCM [14–17]. Karaman et al. [18] used a one-dimensional flow equation to examine the effects of subsidence on water-level by introducing a sink that moves with the mining face and tested the validity of this method by estimating parameters of aquifer over a longwall coal mine in the Illinois. Shultz [19] studied the effects of longwall coal mining on the groundwater hydrology of Marshall County (West Virginia, USA). Booth et al. [20] conducted a seven-year study of a sandstone aquifer overlying an active longwall mine in Illinois (USA) and developed a comprehensive model of its influencing factors. They drew the conclusion that the changes of permeability and storativity over the longwall panel caused the decline and recover of the water levels in the sandstone. Hill et al. [21] conducted a field measurement by placing groundwater monitoring networks over the longwall panel. They collected, compiled and analyzed the data to provide documentation of the groundwater fluctuations caused by mining to local officers. Robertson [22] used Ostrom's design principles for common pool resources and revealed several management challenges of the key aquifers' water-level continued declining in the Great Artesian Basin (Australia). Howladar [23] studied the impact of underground coal mining on water environment around the coal mining area in Dinajpur (India). By analyzing the water level data and ground water major parameters collected from 2001 to 2011, he concluded that without a sustainable and long-term coal mining plan in the area, the water level will deplete and the water crisis will occur in the coming day. Gandhe et al. [24] optimized the mining methods and investigated hydrogeology conditions to protect the surface water bodies from being disturbed or destroyed in Godavari (India).

Fan et al. [25] analyzed the relationship between groundwater table decline and mining intensity in the Yu-Shen mining area based on statistical data of local groundwater table before and after large-scale coal mining. It was found that intensive mining is the primary factor causing the lowering of groundwater table in this area. Based on analysis of the spatial relationship between coal seam, aquifer and aquiclude, Wang et al. [26] divided the study area into several zones according to aquiclude stability after mining and then assessed the applicability of WCCM to different zones. Miao et al. [27] proposed the concept of water-resisting key strata based on the strata control theory. They also analyzed the mechanical behavior of compound key strata using six different composite beam models. Through stability analysis of water-proof strata overlying a shallow coal seam during mining, Huang et al. [28] investigated the fracture development characteristics above and below the water-resisting strata, and proposed the use of water-resisting strata thickness to mining height ratio as the criterion of water resisting capacity assessment. Zhang et al. [29] identified different types of eco-geological environment in the coal mining areas across Northwest China and analyzed the distribution of mining-induced fractures in aquiclude. Based on the stratigraphic structures in typical coal mines of Northwest China, a technological system of WCCM, taking into account the relationship between bedrock thickness, water-bearing of aquifer, mining methods and parameters, was established. Moreover, Ma [2,12,30–33] et al. investigated the development of water conductive fractures induced by shallow seam mining and proposed two WCCM methods. One is applicable for shortwall mining with "mining while filling", and the other is for longwall mining with partial backfilling.

After years of research and practice, WCCM theory and technology have been well-developed and successfully applied in many coal mines of North China and other parts of the world [13,34–37]. However, due to the heavy dependence of WCCM on site specific geological and hydrogeological conditions, a systematic research approach to achieving WCCM in different mining regions is still needed. In this paper, we analyzed the factors affecting WCCM in Yu-Shen mining area (part of Shaanbei coal production base), and then proposed a systematic WCCM method based on "five maps, three zones, and two zoning plans", which would systematize and standardize WCCM research and practice, providing guidance on how to achieve WCCM in Northwest China. This methodology can then be used to develop WCCM in other coal mining regions in other countries besides China.

2. Identification of Factors Influencing WCCM

Previous researchers have investigated factors influencing WCCM from various perspectives by analyzing the relationship between the decline of groundwater table and mining intensity. Adhikary et al. [6] found that the height of the water conductive fractured zone (WCFZ) depended greatly on the working face length, mining height and face advance rate. Wang et al. [26] categorized the spatial relationship between coal seam and aquifer (or aquiclude) into different types based on the thicknesses of overburden, aquifer, and aquiclude. Miao et al. [27] discussed the water-resisting property of compound key strata in overburden with different stratigraphic structures. Huang et al. [28] demonstrated that mining depth, mining height, and the thickness and properties of overburden rock strata are primary factors determining the development of upward and downward fractures in water-resisting strata. Zhang et al. [29] suggested that shallow burial depth and large mining height would result in the greater possibility of both surface water and shallow groundwater loss. In addition, they illustrated the synergistic failure characteristics of aquicludes and adjoining strata from the perspectives of overburden thickness, lithology, and stratigraphic structure. Ma et al. [30–32] categorized the thickness, permeability coefficient and groundwater recharge condition of loose aquifers and proposed an empirical equation to estimate the height of WCFZ.

With different results, these studies focused on various influencing factors, but none of them were comprehensive and systematic. Therefore, this study divided the factors influencing WCCM into three broad categories: engineering geology, hydrogeology and mining method. We then proposed a multilevel model for the comprehensive assessment of these factors, which is summarized in Table 1.

The importance of each influencing factor was assessed using analytic hierarchy process (AHP). The specific steps of calculating the weights of influencing factors are as follows:

Step 1: Influencing factors determination

$$U = \{u_1, u_2, \ldots u_n\} \tag{1}$$

where U is the domain of WCCM; u_1, u_2, $\ldots u_n$ are the influencing factors.

Step 2: Judgment matrix construction

Taking the relative weight evaluation of sub-factors of primary factors of engineering and geological conditions by an expert for example, the judgment matrix are as follows:

$$W_{B_1 \sim C} = \begin{bmatrix} 1 & 1 & 1 & 3 & 5 \\ 1 & 1 & 1 & 3 & 5 \\ 1 & 1 & 1 & 3 & 5 \\ 1/3 & 1/3 & 1/3 & 1 & 5 \\ 1/5 & 1/5 & 1/5 & 1/5 & 1 \end{bmatrix} \tag{2}$$

where W is the comparison discriminant matrix.

Step 3: The largest eigenvalue and corresponding eigenvector calculation

The largest eigenvalue λ_{max} of the matrix is 5.154 and the corresponding eigenvector W = (0.276, 0.276, 0.276, 0.124, 0.047).

Step 4: The consistency test

The consistency was checked by $CR = CI/RI$, where CR is the consistency ratio, RI is average consistency index and CI is the consistency indicator, which is defined as $CI = (\lambda_{max} - n)/(n - 1)$. If $CR < 0.1$, the relative weights are reasonable; otherwise, the judgment matrix needs to be adjusted by redistributing the weights of influencing factors. The CR of the matrix is $0.035 < 0.1$, indicating the weights distribution is reasonable. The weight of engineering and geological conditions is 0.347, then $0.347 \times W$ = (0.096, 0.096, 0.096, 0.043, 0.016), which is the final weight of the five secondary factors among the primary factors given by the expert.

Many researchers, experts and scholars engaged in WCCM have been invited to evaluate the importance of each influencing factor in the AHP model. The mean values of calculation results of their evaluation are listed in Table 1.

As shown in Table 1, the overburden thickness and type of stratigraphic structure have greater impacts than other engineering geological factors. Among the hydrogeological factors, aquiclude thickness has the greatest impact, followed by aquifer thickness and permeability coefficient. In the category of mining method, the impact of effective mining height is an order of magnitude greater than that of other factors; it is therefore considered as the most important factor in mining method.

Table 1. Influencing factors and the weights of water-conserving mining.

Compound Factor Weight B		Sub-Factor Weight C		Sub-Factor Weight D		Sub-Factor Weight E	
1. Engineering and geological conditions B1	0.347	Overburden thickness C1	0.096		0.096		0.096
		Lithology C2	0.070		0.070		0.070
		Physical and mechanical properties C3	0.080		0.080		0.080
		Stratigraphic structure C4	0.084		0.084		0.084
		Geological structure C5	0.017	Faults D1	0.013	① Fault density E1	0.006
						② Fault length E2	0.003
						③ Fault throw index E3	0.004
				Folds D2	0.004	① Fold section coefficient E4	0.0022
						② Fold plane coefficient E5	0.0018
2. Hydrogeological conditions B2	0.539	Aquifer C6	0.202	Recharge D3	0.082		0.082
				Thickness D4	0.040		0.040
				Permeability coefficient D5	0.039		0.039
				Groundwater depth D6	0.016		0.016
				Vertically adjacent aquifers water difference D7	0.025		0.025
		Aquiclude C7	0.337	Thickness D8	0.179		0.179
				Horizon D9	0.031		0.031
				(3) Water-physical properties D10	0.127	① Permeability E6	0.071
						② Expansibility E7	0.033
						③ Water sensibilityE8	0.023
3. Mining method B3	0.114	Mining methods C8	0.019		0.019		0.019
		Mining parameters C9	0.095	Effective mining height D11	0.067		0.067
				Working face inclination D12	0.0075		0.0075
				Mining depth D13	0.0085		0.0085
				Advance velocity D14	0.003		0.003
				Working face dimensions D15	0.003	① Inclination length E9	0.002
						② Strike length E10	0.001
				Coal pillar stability D16	0.006	① Coal pillar sizes E11	0.0032
						② Coal mechanical properties E12	0.0028
Total	1.0	Total	1.000	Total	1.0000	Total	1.0000

3. Overview of Yu-Shen Mining Area and Determination of "Five Maps"

3.1. Geomorphic Features

The Yu-Shen mining area is located the north of Yulin City and is the center of the Jurassic coalfield in north Shaanxi. This area contains proven coal reserves greater than 30 billion tons. The area is located between the Mu Us Desert (southeast of Inner Mongolia) and the Loess Plateau within northern of Shaanxi Province, and characterized by eolian sand in the west and loess ridges and hills in the east as shown in Figure 1.

The overall surface elevation varies between 1200–1300 m, and is higher in the northwest and lower in the southeast. This area fits into the moderate-temperate zone with a semi-arid continental monsoon climate. Local water resource is limited and unevenly distributed both spatially and temporally. The average annual rainfall is only about 400 mm, over 70% of which occurs from July to September, while the average annual evaporation exceeds 1900 mm [38]. Moreover, the ecological environment of this area is vulnerable to mining, and the early development of coal resources has caused a series of geological and ecological problems [12,25].

3.2. Geological Characteristics

The typical stratigraphic column of Yu-Shen mining area and the lithological characteristics are shown in Figure 2. The surface is primarily covered by the Quaternary and Neogene strata, and bedrock outcrops are sporadically distributed in valleys. From oldest to youngest, the strata found in this area include: the Lower Jurassic Fuxian Formation (J_1f), the Yan'an (J_2y), Zhiluo (J_2z), and Anding Formations (J_2a) of Middle Jurassic age, and the Lower Cretaceous Luohe Formation (K_1l), the Neogene System (N), and the Quaternary System (Q) [26–29,31–37].

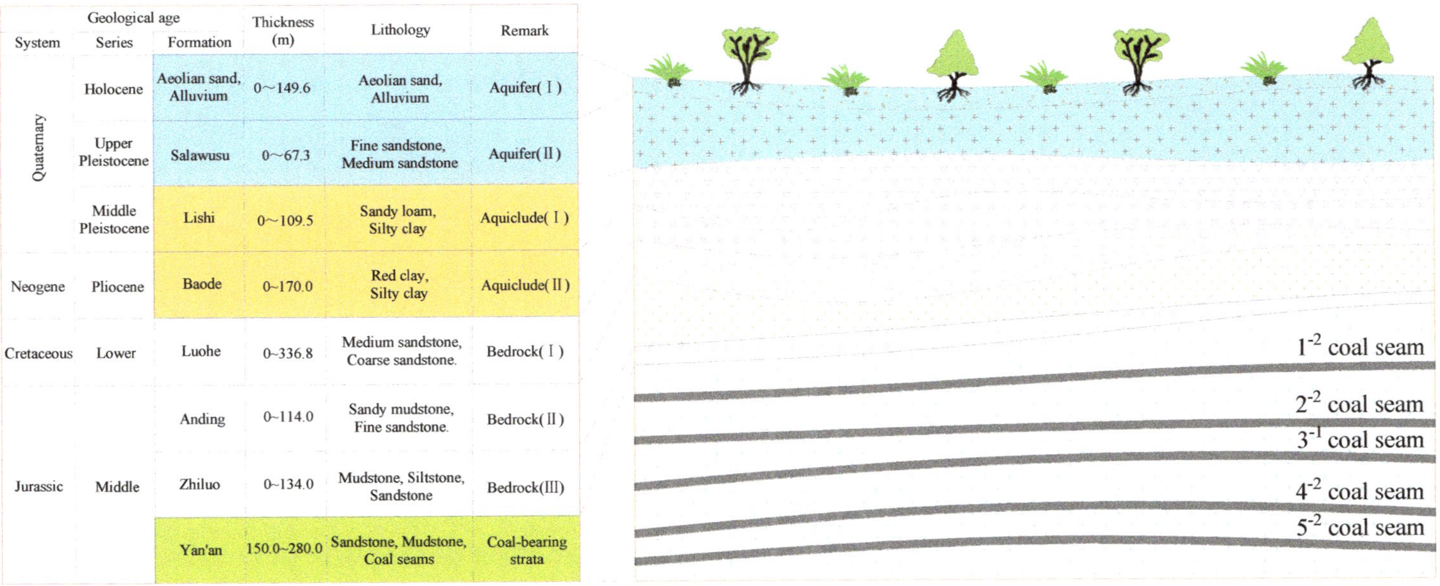

System	Geological age		Thickness (m)	Lithology	Remark
	Series	Formation			
Quaternary	Holocene	Aeolian sand, Alluvium	0~149.6	Aeolian sand, Alluvium	Aquifer(Ⅰ)
	Upper Pleistocene	Salawusu	0~67.3	Fine sandstone, Medium sandstone	Aquifer(Ⅱ)
	Middle Pleistocene	Lishi	0~109.5	Sandy loam, Silty clay	Aquiclude(Ⅰ)
Neogene	Pliocene	Baode	0~170.0	Red clay, Silty clay	Aquiclude(Ⅱ)
Cretaceous	Lower	Luohe	0~336.8	Medium sandstone, Coarse sandstone.	Bedrock(Ⅰ)
Jurassic	Middle	Anding	0~114.0	Sandy mudstone, Fine sandstone.	Bedrock(Ⅱ)
		Zhiluo	0~134.0	Mudstone, Siltstone, Sandstone	Bedrock(Ⅲ)
		Yan'an	150.0~280.0	Sandstone, Mudstone, Coal seams	Coal-bearing strata

Figure 2. Stratigraphic column in Yu-Shen mining area.

3.3. Determination of "Five Maps"

Five maps were generated for the Yu-Shen mining area based on the relative weights (impacts) presented in Table 1, including: overburden thickness contour, stratigraphic structure map, water-rich zoning map of aquifers, aquiclude thickness contour, and coal thickness contour. These maps were used as the input basic parameters for WCCM.

(1) Overburden thickness contour

The overburden strata of Yu-Shen mining area are dominantly sandstone, mudstone, and sandy mudstone. The overburden thickness ranges generally from 0 m to 650 m. As a result of intense weathering and erosion, the overburden gradually decreases from northwest to southeast as shown in Figure 3. The roof strata of the coal seams are primarily composed of medium-hard rocks (compressive strength: 20–40 MPa), such as mudstone with an average UCS (uniaxial compressive strength) of 31.4 MPa and sandstone having an average UCS of 28.9 MPa.

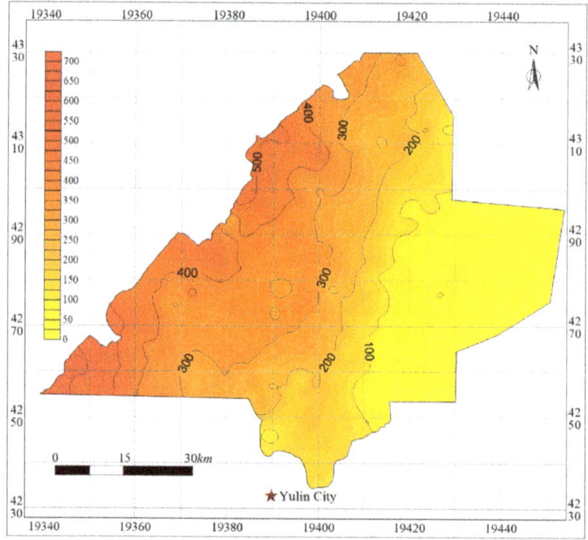

Figure 3. Overburden thickness contour (Xi'an geodetic coordinate system 1980; in km).

The bedrock is weathered and fractured in the first 30 m depth from surface, and the degree of weathering gradually decreases with the increase of depth. The bedrock from 6 m to 10 m depth is highly weathered and secondary structural planes and fracture network was developed. This bedrock layer therefore bears more water than bedrock at high depth. Water yield property of the weathered bedrock varies significantly. It is generally low, except for localized medium levels.

(2) Stratigraphic structure map

The spatial relationship between coal seams and aquifers (or aquicludes) in the Yu-Shen mining area has the following characteristics: The groundwater is found in close association with the underlying coal, and the bedrock is thick and often covered with a thin layer of sandy soil. The distance between coal seam and aquifer varies significantly. The thickness of bedrock above coal seam is negatively correlated with aquifer thickness. A decrease in bedrock thickness indicates increase in aquifer thickness and water yield. An increase in bedrock thickness is associated with decreases in aquifer thickness and lower water yield. The increase in bedrock thickness occurs mostly near boundaries between zones of different water yields.

The stratigraphic structure of overburden in the Yu-Shen mining area was categorized into five types: sand-soil-bedrock (I), sand-bedrock (II), bedrock (III), soil-bedrock (IV), and burnt rock (V) as shown in Figure 4. The overburden of sand-soil-bedrock is composed of sand, impermeable soil and bedrock. This stratigraphic structure is widely distributed in this mining area, covering more than 80% of its territory. The sand-bedrock overburden is very thin. The high-yield unconsolidated aquifer in this structure lies directly above the coal-bearing strata, and there is no continuous clay aquiclude separating them. The bedrock overburden consists of bedrock only which is exposed to the earth's surface. There is no need to consider water conservation here due to the extremely low water bearing capacity. Soil-bedrock overburden is composed of soil and bedrock. As it bears no water, water conservation is not accounted for in this structure. Burnt rock is the product of rock alteration resulting from the spontaneous combustion of coal seams 2^{-2} and 3^{-1}. Burnt rock overburden thickness ranges from 30 m to 50m, and its degree of metamorphism gradually decreases as the increase of vertical distance from the spontaneous combustion seams.

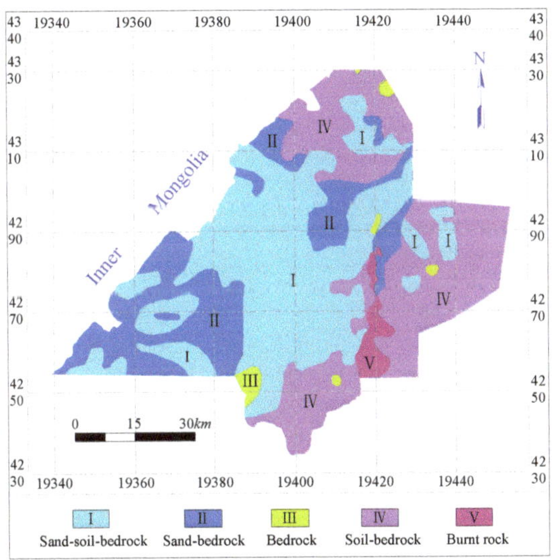

Figure 4. Stratigraphic structure map (Xi'an geodetic coordinate system 1980; in km).

(3) Water-rich zoning map of aquifers

The major types of aquifers in the Yu-Shen mining area, from top to bottom, include: the unconsolidated porous phreatic aquifer, Salawusu Formation aquifer, burnt rock phreatic aquifer, and fractured (porous) bedrock confined aquifer, as shown in Figure 5. The fractured (porous) bedrock confined aquifer has extremely low water content due to limited distribution and relatively high permeability, the burnt rock phreatic aquifer cannot form a water-storing structure. However, it is primarily recharged by groundwater from the water-bearing Salawusu Formation. Therefore, the Salawusu Formation is the major aquifer in this area which needs to be protected during mining.

Holocene (Q_4) porous phreatic aquifers include Holocene eolian sand layer (Q_4^{eol}) and Holocene alluvial deposit (Q_4^{al}). Holocene eolian sand layer (Q_4^{eol}) is widely distributed in Yu-Shen mining area, with an average thickness of 5 m, and is mainly composed of fine sand and silt, this layer is permeable and bears no water. Holocene alluvial deposit (Q_4^{al}) is mainly found in large river valleys and fluvial terraces, with limited distribution and having small thicknesses. This layer basically plays no part in local water supply. These two layers and the underlying Salawusu Formation usually form a complete aquifer.

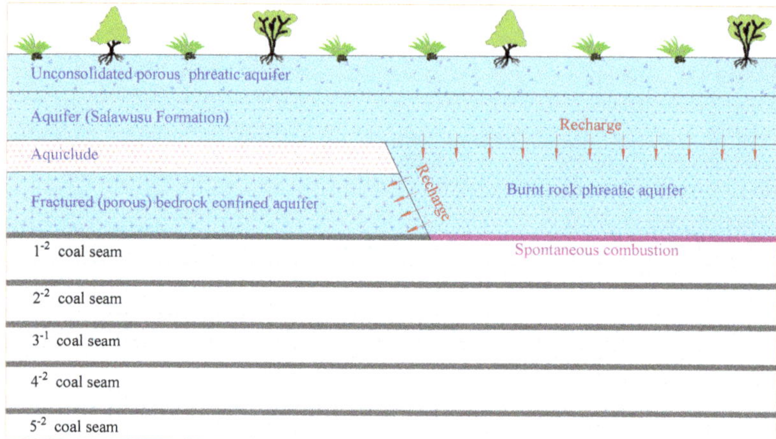

Figure 5. The major types of aquifers in the Yu-Shen mining area.

The water-bearing Salawusu Formation has wide distribution in the mine area. It is normally covered with eolian sand. Its thickness ranges from 0 m to 67.3 m. The water table is generally less than 10 m underground (normally 0.9–9.3 m deep). This formation mainly consists of silt, fine sand, medium sand, and contains sandy loam. With high porosity, the Salawusu Formation is the major aquifer in the study area. Its spatial distribution, thickness, and aquifer structure are strictly governed by modern topographic features and the paleogeographic environment in which it was deposited. The hydrological conditions and water yield of this aquifer vary significantly across this area. The high water-rich part of the aquifer covers a total area of 504 km^2 with a thickness of 39.6–156.1 m and is characterized with permeability coefficient of 0.055–23.6 m/d. Its medium water-rich portion is about 1911 km^2 in total, having depth varies from 6.42–145.50 m and permeability coefficient varies from 0.005–10.457 m/d. The low water-rich portion is 1919 km^2 in total, with the depth varies between 4.4–93.8 m, and permeability coefficient varies between 0.012–6.073 m/d. The water-rich zoning map of aquifers is shown in Figure 6.

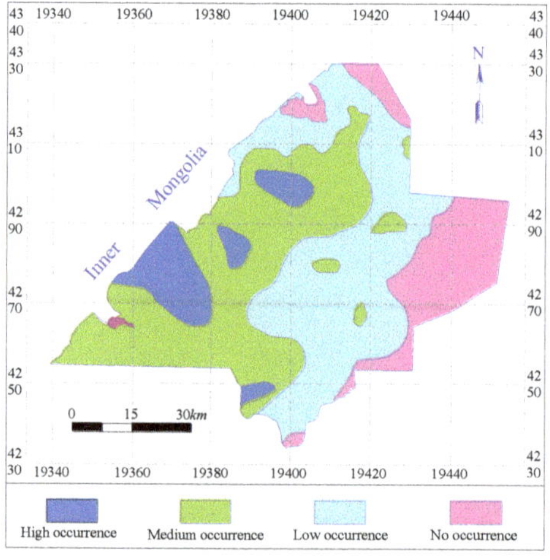

Figure 6. Water-rich zoning map of aquifers (Xi'an geodetic coordinate system 1980; in km).

(4) Aquiclude thickness contour

The main aquicludes in this mining area include the loess in the Lishi Formation and red soil in the Pliocene Baode Formation, as shown in Figure 7. The Lishi Formation is a loess formation composed of grayish-yellow sandy loam and silty clay, with 45.7% sand, 39.5% silt, and 14.8% clay. Its upper part is eolian loess (Q_3m) with well-developed joints and pores. The Lishi loess is well graded, and its coefficient of non-uniformity is 2.0, and the coefficient of curvature is 1.5. The loess's moisture content varies between 11.9–17.3%, plastic limit between 16.9–18.7%, liquid limit between 25.9–31.8%, and plasticity index between 7.9–13.1. These parameters indicate that the loess is in a hard to hard plastic state. The thickness of the Lishi Formation varies with an average of 23.0 m. It is in discordant contact with the underlying layer.

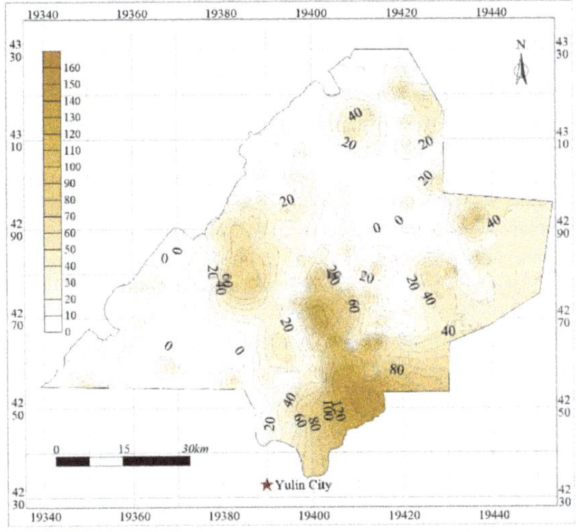

Figure 7. Aquiclude thickness contour map (Xi'an geodetic coordinate system 1980; in km).

The Baode Formation is a red soil formation consisting of brown-red clay and silty clay. The mineralogical composition is chlorite as the dominant mineral of this formation, while kaolinite and illite present in small quantities, having medium plasticity index (from 9.5 to 12.1), medium liquid limit, and low liquidity index, this red soil is a hard plastic clay soil. The thickness of this formation is 30 m on average, which is controlled by the ancient base levels of erosion in this area. It is relatively thick along the water-shed and then decreases gradually towards the valleys on each side.

(5) Coal thickness contour

The Jurassic Yan'an Formation is the coal-bearing formation in this area. It contains five main coal seams referred to as 1^{-2}, 2^{-2}, 3^{-1}, 4^{-2} and 5^{-2}. These seams are generally flat with small dip angles of 1° to 3° but their thicknesses vary significantly. The maximum seam thickness is greater than 12 m, as shown in Figure 8. Seam 1^{-2} occurs in the upper part of the 5th section of the Yan'an Formation. Seams 2^{-2} and 3^{-1} are at the top of the 4th and 3rd sections, respectively. Seam 4^{-2} is in the mid-upper part of the 2nd section of this formation and 5^{-2} is at the top of the first section.

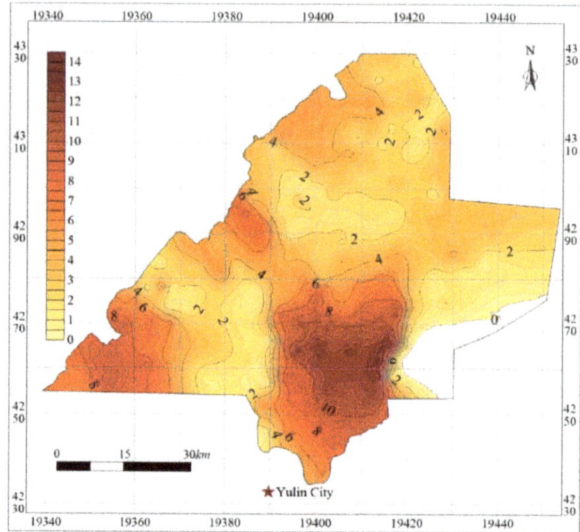

Figure 8. Coal thickness contour (Xi'an geodetic coordinate system 1980; in km).

4. Determination of the Heights of "Three Zones"

4.1. Longwall Mining

In a longwall mine, several interconnected roadways are developed as passageways (often called gate roads) for operators, equipment, coal transport, and ventilation. These roadways split the coal seam into longwall panels, which are typically 100–300 m wide and 1000–5000 m long. A shearer is mounted along the width of each panel, i.e., the longwall face, and cuts coal in 0.6–1.2 m thick slices from the seam as it moves along the face. Chain pillars are usually left between panels, and the roadways on the two sides of a pillar are connected by crosscuts as shown in Figure 9.

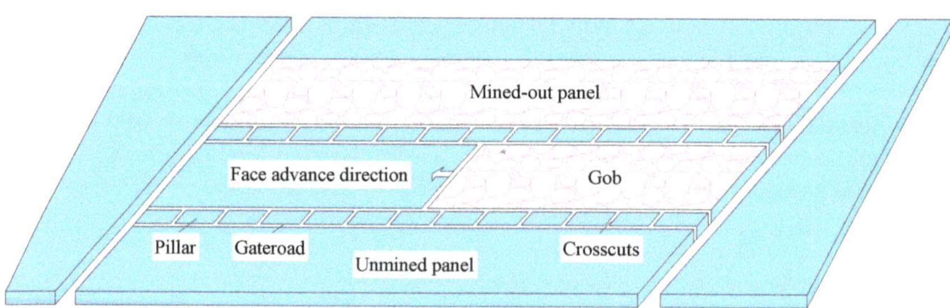

Figure 9. Longwall panel layout.

Longwall mining has been widely used in the United States, Australia, Poland and India because of its significant advantages, e.g., continuous mining, high productivity, high efficiency and high seam recovery rate etc. It is also extensively used by coal mines in Northwest China, where the coal seams are typically thick and shallow [25]. However, high intensity longwall mining leads to severe fracturing in the overlying strata to reach the aquifer, which cause groundwater to infiltrate to longwall working face along the fractures and surface vegetation to die, as shown in Figure 10.

After the coal seam is extracted, the overburden tends to collapse, forming goaf (gob) behind face. Generally, the goaf can be divided into three zones based on the degree of damage: caved zone, fractured zone, and continuous deformation zone as shown in Figure 10 [39,40]. The caved zone and fractured zone are known as the WCFZ as they provide the main paths for groundwater flow [41].

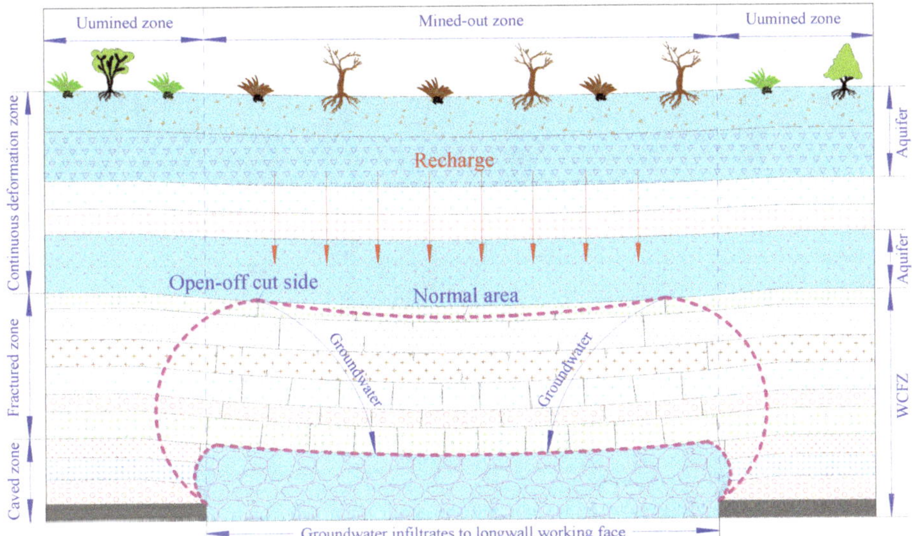

Figure 10. The three zones in overlying strata and groundwater inrush resulting from longwall mining.

4.2. Height of Caved Zone

The ratio of caved zone height to mining height in Yu-Shen mining area was measured and found between 4.12 and 6.38 as describe in Table 2. According to the measured data, the equation of calculating the height of caved zone is obtained:

$$H_k = \frac{100M}{0.6M + 14.1} \tag{3}$$

where, H_k is the height of caved zone (m) and M is the effective mining height (m).

Table 2. Field observation of caved zone height and WCFZ height in Yu-Shen mining area.

Coal Mine	Working Face/Borehole	Mining Height (m)	Caved Zone Height (m)	The Ratio of Caved Zone Height to Mining Height	WCFZ Height (m)	The Ratio of WCFZ Height to Mining Height
Yuyang	2304	3.5	17.2	4.91	96.3	27.51
Jinjitan	101	5.5	23.14	4.2	108.59	20.54
	Y3	5	24.50	4.90	130.50	26.10
	Y4	5	27.90	5.58	137.30	27.46
Yushuwan	Y5	5	27.10	5.42	138.90	27.78
	Y6	5	20.60	4.12	117.80	23.56
	H3	4.5	20.50	4.56	108.32	24.07
	H4	4.5	22.18	4.93	114.38	25.42
Hanglaiwan	H5	4.5	19.40	4.31	107.83	23.96
	H6	4.5	28.70	6.38	93.87	20.86

The height of the caved zone near each borehole was calculated using Equation (3). The results obtained is plotted showing the caved zone height in an area as shown in Figure 11.

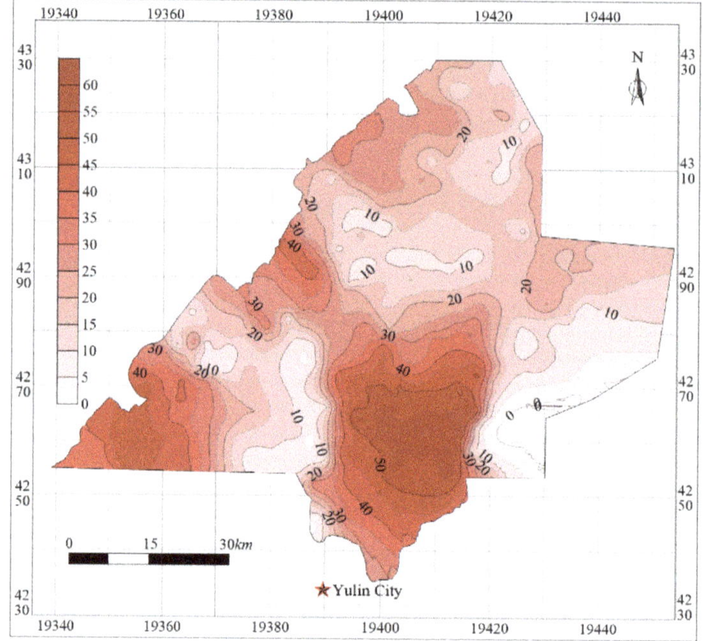

Figure 11. Caved zone height contour (Xi'an geodetic coordinate system 1980; in km).

4.3. Height of WCFZ

To determine the height of WCFZ, four main methods can be used: field measurement, numerical modeling, empirical equation, and physical modeling. As it is difficult to obtain accurate results by each single method, a combination of two or more of them is usually used.

(1) Field measurements

The location of top boundary and height of WCFZ induced by mining of each main seam in Yu-Shen mining area were determined through observation of: drilling fluid loss combined with engineering geological logging of rock cores, packer testing, borehole television logging, and geophysical logging [42,43]. The measured heights of WCFZ detail are given in Table 2.

The average measured ratio of WCFZ height to mining height in the study area was 24.73. Overall, the ratio decreases with the increase of mining height.

(2) Numerical calculation

The thickness of the main coal seam in the Yu-Shen mining area varies between 0.3–12.4 m, while the mining height obtained according to the measured height of WCFZ varies from 3.5 to 5.5 m. In order to thoroughly analyze the height of WCFZ under different stratigraphic structures and mining heights, the universal distinct element code (UDEC) was employed and four numerical calculation models were developed. In these numerical models, the height of WCFZ under different mining heights was calculated for stratigraphic structures of sand-soil-bedrock, sand-bedrock, bedrock and soil-bedrock respectively. It is noted that, for the stratigraphic structure of burnt rock, the burial depth of coal seam is less than 150 m, and the mining-induced WCFZ will generally develop and reach the surface. Therefore, it is not necessary to simulate the height of the WCFZ of the burnt rock.

Based on the stratigraphic column of the Yu-Shen mining area, numerical calculation models were developed to simulate the height of WCFZ based on different boreholes. The model parameters were continuously optimized until a good agreement of the numerical results and the field measured results was reached. Taking the typical sand-soil-bedrock stratigraphic structure as example, similar lithologic strata were merged and 14 strata were identified from bottom to top according to the Y5 borehole in Yushuwan Coal Mine. The numerical model with dimensions of 800 m × 302 m was constructed based on reducing the boundary effect and supercritical mining as shown in Figure 12. The left and right sides of the model were fixed in the X direction, and the bottom boundary of the model was fixed in the Y direction. Mohr-Coulomb model was used during the calculation. The WCFZ height for 5 m mining height is calculated and continuously adjusting parameters according to the numerical results, the optimum model parameters which can result in the measured results were obtained, and they are shown in Tables 3 and 4.

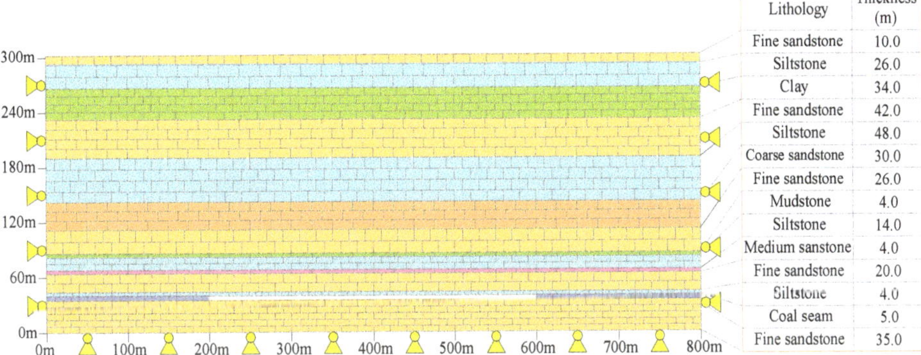

Figure 12. Numerical simulation model of sand-soil-bedrock stratigraphic structure.

Table 3. Rock physical and mechanical parameters (block).

Number	Strata	Thickness (m)	Density (kg/m^3)	Bulk Modulus (GPa)	Shear Modulus (GPa)	Cohesion (MPa)	Friction Angle (°)	Tensile Strength (MPa)
1	Fine sandstone	10	2600	30.8	20.3	5.6	35	4.0
2	Siltstone	26	2460	16.1	11.6	2.0	21	1.2
3	Clay	34	1900	0.28	0.093	0.85	25	0.35
4	Coarse sandstone	30	2500	20.8	11.9	3.0	23	1.4
5	Mudstone	4	2200	8.3	4.3	2.1	25	1.0
6	Medium sandstone	4	2560	23.1	14.5	4.4	2.8	2.0
7	Coal seam	10	1400	2.0	1.4	1.7	28	1.5

Table 4. Rock physical and mechanical parameters (contact).

Number	Strata	Normal Stiffness (GPa)	Shear Stiffness (GPa)	Cohesion (MPa)	Friction Angle (°)	Tensile Strength (MPa)
1	Fine sandstone	300	200	6.5	12	3.5
2	Siltstone	200	160	8.0	10	5.5
3	Clay	300	180	6.0	20	3.0
4	Coarse sandstone	500	400	4.0	25	3.0
5	Mudstone	300	100	2.1	20	1.5
6	Medium sandstone	150	100	4.2	15	2.2
7	Coal seam	500	350	9.0	15	6.0

The WCFZ of 5 m mining height is shown in Figure 13. The optimized model parameters were used, the heights of WCFZ zone under different mining heights were calculated for 4 different stratigraphic structures, and the results are shown in Table 5.

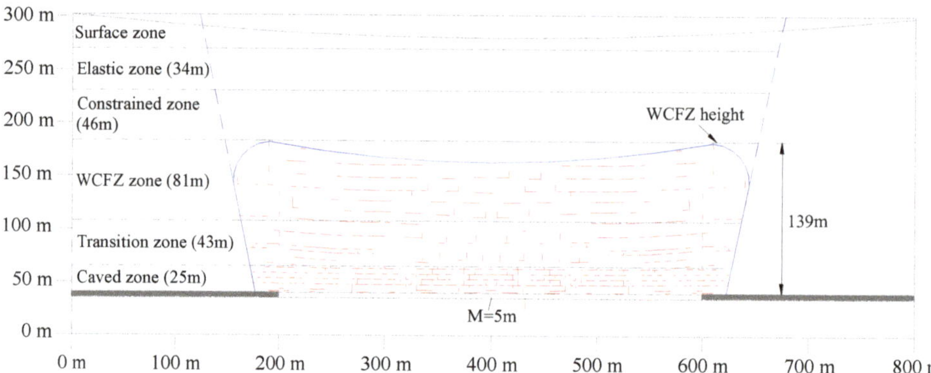

Figure 13. WCFZ of 5 m mining height of sand-soil-bedrock stratigraphic structure.

Table 5. Numerical simulation results of WCFZ height and the ratio of WCFZ height to mining height.

Mining Height (m)	Sand-Soil-Bedrock		Sand-Bedrock		Bedrock		Soil-Bedrock	
	WCFZ Height (m)	The Ratio of WCFZ Height to Mining Height	WCFZ Height (m)	The Ratio of WCFZ Height to Mining Height	WCFZ Height (m)	The Ratio of WCFZ Height to Mining Height	WCFZ Height (m)	The Ratio of WCFZ Height to Mining Height
2	64	32.0	68	34.0	53	26.5	67	33.5
3	92	30.7	101	33.7	78	26.0	91	30.3
4	119	29.8	131	32.8	101	25.3	113	28.3
5	139	27.8	154	30.8	122	24.4	141	28.2
6	165	27.5	179	29.8	138	23.0	169	28.2
7	184	26.3	200	28.6	153	21.9	190	27.1
8	208	26.0	221	27.6	170	21.3	211	26.4
9	218	24.2	235	26.1	177	19.7	226	25.1
10	240	24.0	244	24.4	187	18.7	234	23.4

(3) Calculation equation

The equation for WCFZ height calculation was obtained by regression analysis using field measured data under medium thickness seam mining condition in eastern China [31,32]. However, due to the change of coal occurrence, geological and mining conditions, the conventional equation is not suitable for use in the Yu-Shen mining area [44]. Therefore, it is necessary to derive an equation which suits the regional geology based on the measured heights of WCFZ combined with the numerical results [30,33,45–47].

Based on field observations and the numerical simulation results shown in Tables 2 and 5 respectively, four equations for predicting the height of WCFZ were obtained by regression analysis using different stratigraphic structures in Yu-Shen mining area:

$$\begin{cases} H_d = 21.75M + 28.28 & R^2 = 0.99 \quad (\text{Sand} - \text{soil} - \text{bedrock}) \\ H_d = 22.2M + 37.13 & R^2 = 0.97 \quad (\text{Sand} - \text{bedrock}) \\ H_d = 16.7M + 30.8 & R^2 = 0.97 \quad (\text{Bedrock}) \\ H_d = 21.97M + 28.42 & R^2 = 0.98 \quad (\text{Soil} - \text{bedrock}) \end{cases} \tag{4}$$

where, H_d is the height of WCFZ; M is the mining height.

Using Equation (4), the heights of WCFZ at different borehole positions were calculated. The WCFZ height is shown in Figure 14.

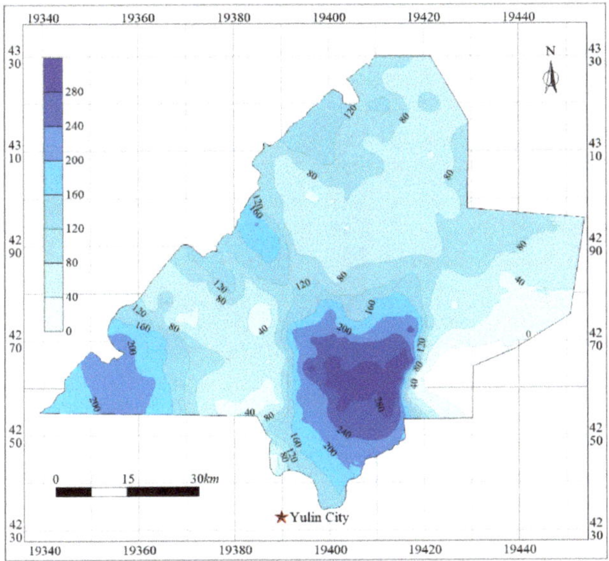

Figure 14. WCFZ height contour (Xi'an geodetic coordinate system 1980, in km).

4.4. Thickness of Protective Zone

A protective zone overlying the WCFZ can help to prevent water in the overlying aquifer from flowing into coal faces [31,32]. A water-resisting index based on rock strength was proposed for the quantitative assessment of the water-resisting property of this zone [48]. The more impermeable the protective zone, the higher the index and the harder the rock in the protective zone, the lower the index. If there is a continuous clay layer greater than 3 m thick separating the WCFZ from the overlying aquifer, it can effectively prevent the downward percolation of water from the aquifer. The amount of water loss, even if only relatively small can cause considerable ecological damage, so the acceptable the safety factor was set at 4. This means that for a clay layer to be an effective protective zone, its thickness should be at least 12 m. If water-resisting index of clay is 1, the water-resisting index of the bedrock in the Yu-Shen mining area would be 0.4 for bedrock of sandstone, mudstone and sandy mudstone. Therefore, with the absence of clay, the thickness of bedrock should be at least 30 m to be an effective protective zone. The method for determining the thickness of protective zone can be expressed as follows:

$$H_b = \begin{cases} 12 & (H_t \geq 12) \\ 30 - 1.5H_t & (0 \leq H_t < 12) \end{cases} \tag{5}$$

where, H_b is the thickness of protective zone (m) and H_t is thickness of clay layer (m).

Using Equation (5), the thicknesses of protective zone at different boreholes in the Yu-Shen mining area were determined, and the results can be observed in the contour as shown in Figure 15.

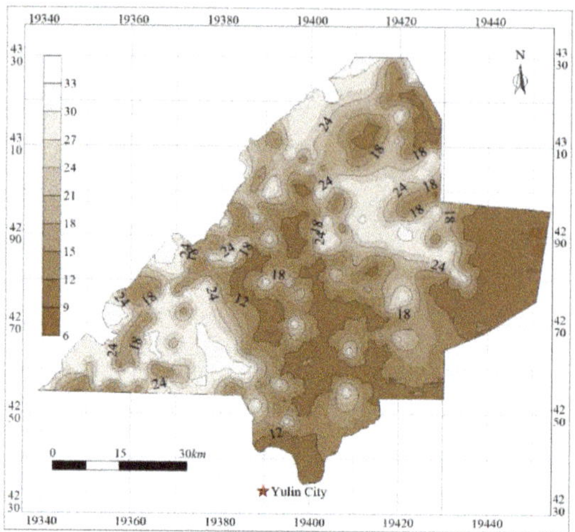

Figure 15. Protection zone thickness contour (Xi'an geodetic coordinate system 1980; in km).

5. Zoning Based on Coal Mining's Impact Level on Groundwater

There are five main seams in the Yu-Shen mining area and their thicknesses vary widely. The overall overburden thickness increases gradually from east to west, but regional increase or decrease occurs as a result of change in sedimentary environment or denudation. As the roof strata are relatively unstable, mining activities can easily induce the development of "three zones" (caved zone, WCFZ and protective zone) in the vertical direction. These factors determine that the loss of aquifer water during mining varies spatially and different WCCM method should be implemented according to local conditions. Based on measured bedrock thickness, the spatial relationship between caved zone and WCFZ induced by mining, and the thickness of protective zone, coal mining's impact level on groundwater was classified using the standard summarized in Table 6. Then the study area was divided into zones of different impact levels according to the results, as illustrated in Figure 16:

(1) Unconfirmed zone: the geological data are unavailable in this zone.

(2) No impact zone: the Salawusu Formation is thin or absent, thus there is no effective aquifer.

(3) Slightly impacted zone: the thickness of bedrock over the first mined seam + soil thickness > the height of WCFZ + protective zone thickness. After mining, the WCFZ and aquifer are separated by continuous strata that are thick and impermeable enough to prevent percolation. The aquifer will be effectively protected from mining-induced water loss.

(4) Moderately impacted zone: the height of WCFZ < the thickness of bedrock over the first mined seam + soil thickness < the height of WCFZ + protective zone thickness. There are continuous strata with certain thickness between the WCFZ and the aquifer, but the strata are unable to completely prevent the downward percolation of water from the aquifer.

(5) 'Severely impacted zone: the height of caved zone + protective zone thickness < the thickness of bedrock over the first mined seam + soil thickness < the height of WCFZ. Mining-induced fractures will extend to the aquifer resulting in groundwater loss from the aquifer. The protective zone could only prevent sand inrush.

(6) 'Extremely impacted zone: the thickness of bedrock over the first mined seam + soil thickness < height of caved zone + protective zone thickness. Coal extraction may cause inrush of water and sand.

Table 6. Classification standard for coal mining's impact level on groundwater.

Extremely Severe Impacted Zone	Severe Impacted Zone	Moderate Impacted Zone	Slight Impacted Zone
$H + H_t < H_k + H_b$	$H_k + H_b < H + H_t < H_d$	$H_d < H + H_t < H_d + H_b$	$H + H_t > H_d + H_b$

Where H refers to the height of bedrock strata, m; H_t refers to the height of soil strata, m; H_k refers to the height of the caved zone, m; H_b refers to the thickness of the protective zone, m; H_d refers to the height of WCFZ, m.

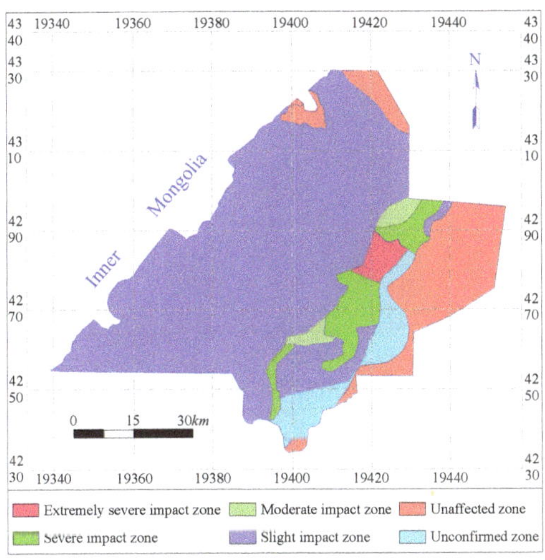

Figure 16. Zoning the extent of coal mining impact on groundwater. (Xi'an geodetic coordinate system 1980; in km).

6. Zoning Based on Applicability of WCCM Methods

A system of WCCM methods mainly including restricted mining height, (partial) backfill, and/or narrow strip mining has been established after years of research and practice [49]. Prior to carrying out field practice under site specific geological conditions, the applicability of these methods needs to be considered. The Yu-Shen mining area was then partitioned into zones that are suitable for different WCCM methods.

6.1. Maximum (Theoretical) Allowable Mining Height

According to the classification standard for coal mining's impact level on groundwater. If the thickness of soil strata plus the thickness of bedrock is greater than the height of WCFZ plus the height of protective zone, then the WCCM can be realized. Subtracting the necessary thickness of protective zone from the total thickness of soil and bedrock yields the height of WCFZ. The theoretical allowable mining height at each longwall face, M_c, can be calculated according to Equation (4). The theoretical allowable mining height obtained was then compared with the actual coal thickness, M_m. Furthermore, the maximum allowable mining height (MAMH), namely the maximum mining height allowed by WCCM, can be derived using Equation (4) under the four different stratigraphic structures:

(1) The equation of MAMH for sand-soil-bedrock overburden is as follows:

$$M_c = \begin{cases} \frac{(H+H_t-12)-28.28}{21.75} & (H_t \geq 12) \\ \frac{[H+H_t-(30-1.5H_t)]-28.28}{21.75} & (0 \leq H_t < 12) \end{cases} \tag{6}$$

(2) The equation of MAMH for sand-bedrock overburden is as follows:

$$M_c = \begin{cases} \dfrac{(H+H_t-12)-37.13}{22.2} & (H_t \geq 12) \\ \dfrac{[H+H_t-(30-1.5H_t)]-37.13}{22.2} & (0 \leq H_t < 12) \end{cases} \tag{7}$$

(3) The equation of MAMH for bedrock overburden is as follows:

$$M_c = \begin{cases} \dfrac{(H+H_t-12)-30.8}{16.7} & (H_t \geq 12) \\ \dfrac{[H+H_t-(30-1.5H_t)]-30.8}{16.7} & (0 \leq H_t < 12) \end{cases} \tag{8}$$

(4) The equation of MAMH for soil-bedrock overburden is as follows:

$$M_c = \begin{cases} \dfrac{(H+H_t-12)-28.42}{21.97} & (H_t \geq 12) \\ \dfrac{[H+H_t-(30-1.5H_t)]-28.42}{21.97} & (0 \leq H_t < 12) \end{cases} \tag{9}$$

where, M_c is the maximum allowable mining height (m); H is bedrock thickness (m); H_t is soil thickness (m).

The MAMH contour calculated from Equations (6) to (9), indicating its distribution across Yu-Shen mining area is shown Figure 17.

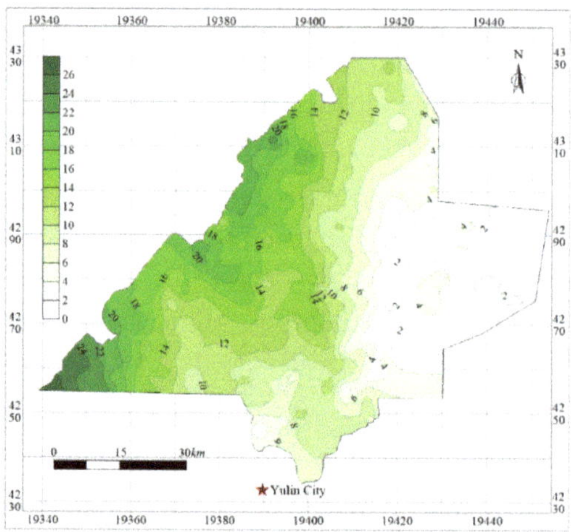

Figure 17. Maximum (theoretical) allowable mining height contour. (Xi'an geodetic coordinate system 1980; in km).

6.2. Height-Restricted Mining

In height-restricted mining areas, the mining height is limited for the purpose of reducing overburden displacement and thereby the height of WCFZ. This method can be easily implemented in thick coal seams without extra cost. In addition, the roof over the goaf will collapse completely, thus avoiding the problem of coal pillar failure as occurred in other methods. However, as the coal seam is partially extracted, the recovery rate is obviously lower when using this method.

In height-restricted mining, WCCM is achieved by reducing the actual mining height. Theoretically, it is applicable throughout the study area. Considering the mine production, profit, mining equipment layout and other relevant factors, this method should be more suitable for zones where the maximum allowable mining height (M_c) is equal or greater than 2 m. The distribution of zones to which height-restricted mining is applicable is shown in Figure 18a.

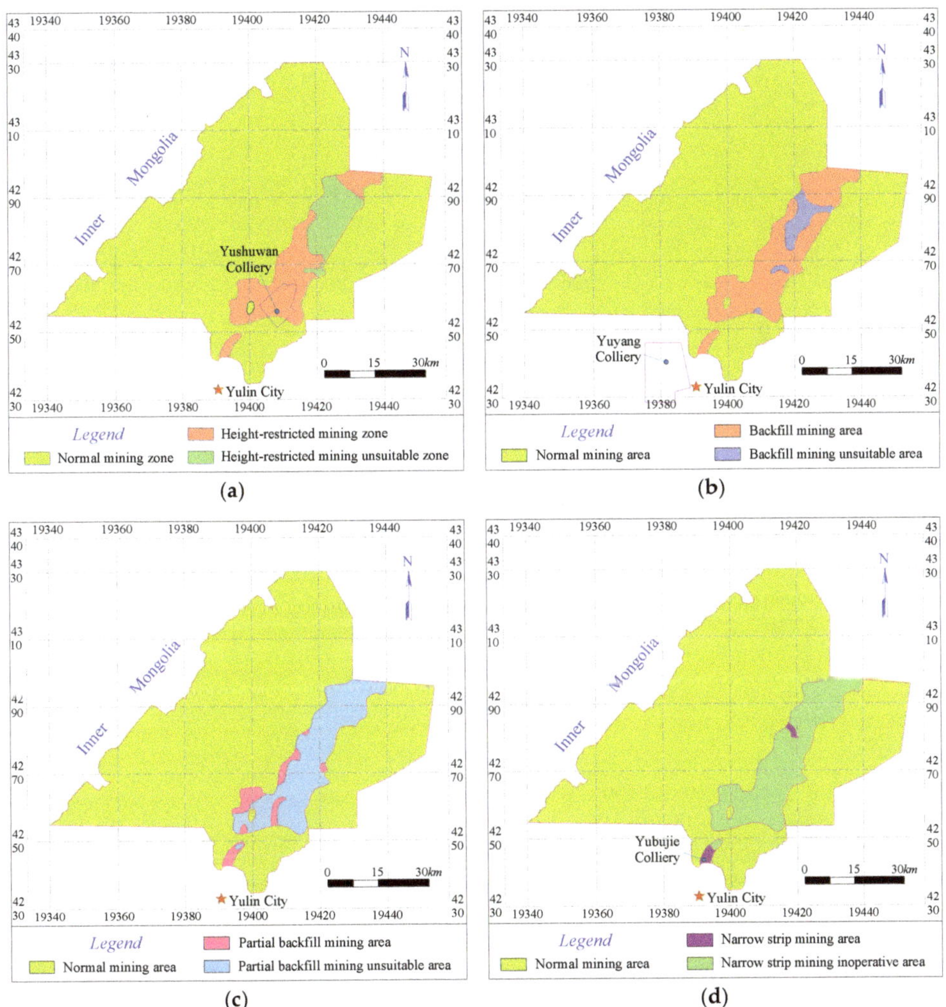

Figure 18. Applicability partition of water conservation coal mining method (Xi'an geodetic coordinate system 1980; in km): (**a**) Height-restricted mining application area; (**b**) Backfill mining application area; (**c**) Partial backfill mining application area; (**d**) Narrow strip mining application area.

6.3. Backfill Mining

In backfill mining, goaf is refilled with backfill concurrently with mining to support the roof, so to control the overburden movement and deformation, the height of WCFZ is therefore reduced. In this way, aquifers are protected from being disturbed by mining activities and to achieve water conservation. In theory, this method allows all coal resources in an area to be extracted. It is recognized as an ideal approach to protecting water resources from coal mining-induced damage and is an important component of green mining technology [50]. However, the disadvantages of backfill mining, such as higher cost and large backfill requirement, have restricted its wide applications in coal mines.

Considering the time required for the filling body to reach the designed strength and the value of roof displacement before working face is filled. It is generally accepted in the mining industry that the maximum backfill mining height is 4.0 m. The inherent compressibility properties of backfill materials, it is impossible to achieve 100% filling ratio. Considering a filling ratio of 75%, backfill mining is applicable to zones where $M_c \geq (1 - 75\%)\, M_m$, and the corresponding minimum backfill height is $(M_m - M_c) / 75\%$. The zones which are suitable for the use of backfill mining in Yu-Shen mining area is shown in Figure 18b.

6.4. Partial Backfill Mining

In the Yu-Shen mining area, the height of WCFZ at normal areas of a longwall face is about 10–15% less than that at the open-off cut and the location of first main roof weighting (after 2 to 3 periodic weighting events), as shown in Figure 10. According to the development characteristics of WCFZ, backfill can be carried out around the open-off cut and near the location of first main roof weighting, so to reduce regional stress concentration in the roof and its damage to overburden. As longwall face advances, the consequence of backfill gradually weakens and the stress in the roof will increase again. The partially backfilling at the two above mentioned positions, the roof would collapse more gently than normal face without backfilling, and the roof movement would be in the overall deformation stage.

Partial backfill is able to effectively reduce the heights of WCFZ of the open-off cut zone and the first weighting zone effectively. However, its effect on controlling the development of water conductive fractures is limited when compared with other WCCM methods. Therefore, this method can be used as a supplement to other WCCM methods or adopted in regions with lower water loss risk, such as areas where the height of WCFZ is greater than bedrock thickness but within 15% of difference.

Therefore, when the sand-soil-bedrock overburden satisfies Equation (10), partial backfill mining should be employed:

$$\frac{21.75(M_m - M_c) + 28.28}{21.75 M_m + 28.28} \leq 15\% \tag{10}$$

When the sand-bedrock overburden satisfies Equation (11), partial backfill mining should be employed:

$$\frac{22.2(M_m - M_c) + 37.13}{22.2 M_m + 37.13} \leq 15\% \tag{11}$$

When the bedrock overburden satisfies Equation (12), partial backfill mining should be employed:

$$\frac{16.7(M_m - M_c) + 30.8}{16.7 M_m + 30.8} \leq 15\% \tag{12}$$

When the soil-bedrock overburden satisfies Equation (13), partial backfill mining should be employed:

$$\frac{21.97(M_m - M_c) + 28.42}{21.97 M_m + 28.42} \leq 15\% \tag{13}$$

where, M_c denotes the maximum allowable mining height (m); M_m is coal thickness (m).

Figure 18c shows the locations suitable for partial backfill mining in Yu-Shen mining area.

6.5. Narrow Strip Mining

In narrow strip mining, the block of coal to be mined is divided into strips. After one strip is extracted, the adjacent strip will be left in place to support the overlying strata and the strip next to it will be extracted (similar to drift and pillar mining), as shown in Figure 19. The overburden will thus displace and deform more mildly and evenly than it in the full-seam extraction, maintaining the structural stability of aquicludes during mining, and thereby protecting water resources. However, this method achieves WCCM at the cost of low recovery rate, which is similar to that of height-restricted mining. This is also the main reason restricting its use for water conservation.

In addition, as the coal pillars left to support overburden may fail due to weathering after years of extraction, narrow strip mining is only suitable under certain geological conditions. The recovery of coal pillars and the associated safety issues within the goaf are still challenging issues when using narrow strip mining. Therefore, narrow strip mining can be adopted at low-production coal mines with thin seams as long as local conditions permit. Taken into account the current technological and economic conditions and equipment available in the Yu-Shen mining area, narrow strip mining should be a suitable WCCM method for mines with annual production below 0.9 Mt/a (million tons every year). Figure 18d shows the zones suitable for this method to use within the Yu-Shen mining area. It is worth noting that room and pillar mining has been employed by some local coal mines for many years. This method has two major weaknesses, low recovery rate and non-standard face layout.

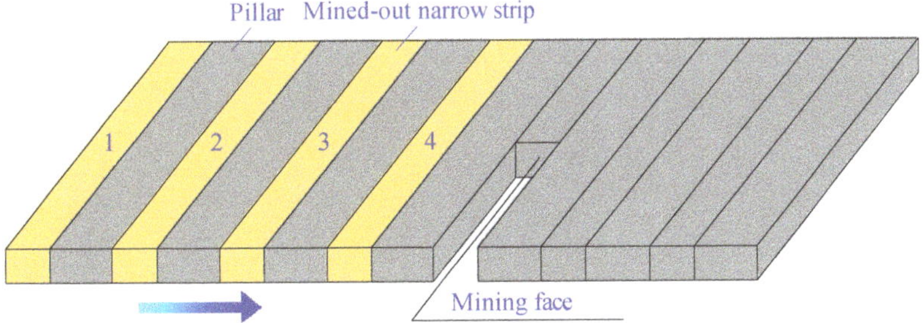

Figure 19. Sketch map of narrow strip mining.

Despite being relatively old-style, narrow strip mining could be an option for some small and medium sized mines to repace the room and pillar mining, whcih can increase recovery rate by over 20% while also protecting groundwater [7]. This would be particularly useful in coal areas where longwall panels are difficult to design due to adverse geology for example.

7. Case Studies

7.1. Height-Restricted Mining

The Yushuwan Coal Mine, located in the south of the study area is shown in Figure 18a. It has five minable coal seams and Seam 2^{-2} is uppermost and 11 m thick. The bedrock thickness varies from 85–142 m and the impermeable soil thickness is around 73–96 m. The mine is categorized into sand-soil-bedrock stratigraphic structure. According to Equation (6), the MAMH was calculated and found between 7–9 m, which is smaller than the actual seam thickness. This indicates that WCCM is needed. Early suggested in Section 6.1, height-restricted mining could be adopted at this mine. Eventually, the actual mining height for the Yushuwan Coal Mine was determined to be 5.5 m. More than five longwall panels have been successfully extracted now with no mining induced WCFZ reaching the overlying Salawusu Formation, demonstrating the achievement of WCCM and local water resource is effectively protected [51].

7.2. Backfill Mining

The main seam being mined at the Yuyang Coal Mine is between 2 m and 4 m thick. The overlying bedrock is 60 m to 310 m thick, and the impermeable soil thickness varies between 15 m and 120 m. Though this mine is within Yuyang District, Yulin City, its local geology is similar to that of sand-soil-bedrock stratigraphic structure of Yu-Shen mining area due to their geographic proximity (Figure 18b). The MAMH for this mine was calculated using Equation (6) to vary from 3 to 16 m. A comparison with the actual coal thickness suggests that WCCM method needs to be implemented in some parts of mine. The Yuyang coal mine takes eolian sand from the surface as a major component of the paste-like backfill material in order to reduce the backfill cost. At shortwall face 2301, about 5.2×10^4 m^3 of backfill material has been injected into 50 branch roadways, through with the overburden movement and failure were effectively controlled. According to measurements conducted after mining, the surface subsidence decreased by over 50% and the water table declined by less than 1.61 m on average, implying good water protection effect has been reached [52,53].

7.3. Narrow Strip Mining

The Yubujie Coal Mine is a small mine (annual production: 0.3 Mt/a) located at the southern boundary of the Yu-Shen mining area (Figure 18d). Seam 2^{-2}, the uppermost minable seam, is 4–6 m thick; the overlying bedrock is 90–100 m thick and impermeable soil is 20–90 m thick. The mine geology is categorized into soil-bedrock stratigraphic structure. Using Equation (9), the MAMH was estimated to vary between 4–6 m, indicating that WCCM method is needed in some parts of this mine. Given the small size of this mine (with annual production below 0.9 Mt/a), narrow strip mining was used. The coal strips mined are 12 m wide and the coal pillars left unmined are 8 m wide. This has been used at face 300 and other faces. Field observations demonstrated that the coal pillars were stable and the roof was general intact. The decline of water level in the Salawusu Formation was less than 0.3 m, and the water table remains within 5 m below surface, demonstrating the effective protection of the local ecological water environment [7,54–56].

8. Discussion

(1) AHP is a subjective weighting method. Since the weight used in the present study was mainly provided by scholars for WCCM in Northwest China, further research is needed to confirm whether the weights determined for different influencing factors are applicable to other mining areas.

(2) The proposed equation for the height of WCFZ in the Yu-Shen mining area was only related to mining height, which is similar to the empirical equation provided in the statutory regulations. According to existing empirical and theoretical research, the height of WCFZ induced by coal mining is also significantly affected by several other factors, e.g., overburden structure, roof management method, face length, and burial depth. The relationship between the height of WCFZ and these factors is nonlinear and hard to quantify. At present, no research has provided satisfactory quantitative results about the mechanisms and levels of their influences on the height of caved and fractured zones. Field production practice suggests that when a face is short (e.g., less than 50 m), the overburden can form a self-balancing arch and face length has limited influence on the height of WCFZ. Under subcritical mining condition, the height of WCFZ tends to increase with the increase of face length. Under critical mining condition, it continues to increase with the increase of face length, but the rate of increase is relatively low, and the vertical extension of this zone will basically terminate at a particular height. In terms of the relationship between the height of WCFZ and mining depth, it is found that in-situ stress rises with the increase of mining depth within a certain mining depth, and the zone of stress-relief horizontal fractures in roof strata will also expand vertically. When mining reaches a certain depth, the height of WCFZ will not increase with the increase of mining depth. This can be attributed to the high horizontal in-situ stress which could force the mining-induced stress-relief fractures to close rapidly. The data used to investigate the impact of advance rate on the height of WCFZ were generally collected two months after the face passing by. As a result, by the time of measurement, the overburden strata displacement had already become stable and the WCFZ had basically stopped developing in the vertical direction. Therefore, the results of the regression analysis based on the measurements could not reflect the actual impact of advance rate on the height of WCFZ. Under the specific geological conditions of the Yu-Shen mining area (including structure of overburden and mining depth) and mining method (including roof management method and face length), the height of WCFZ depends heavily on mining height. In addition, the coefficients in the proposed Equation also take into account the effects of overburden structure, roof management method, face length, and mining depth on WCFZ height. Thus, it is reasonable in study to derive the equation of WCFZ height by regression analysis only based on mining heights. This Equation can provide a scientific basis for WCCM, preventing water percolating from roof in Yu-Shen mining area.

(3) The applicability of different WCCM methods was determined based on the specific geological conditions of the Yu-Shen mining area together with the available equipment and techniques. For example, height-restricted mining is suitable for regions where the currently used mining equipment can still be used even after imposing a height limit on it (MAMH > 2 m). Backfill mining is applicable to zones where the effective mining height (i.e., the mining height minus the height of compacted backfill) satisfies the requirement of WCCM (filling ratio = 75%). Partial backfill mining is applicable to zones where the difference between the actual height of WCFZ and its theoretical value required for water conservation is smaller than 15%. When applied to other mining areas in Northwest China, the criteria for zoning based on applicability of WCCM methods used in this study should be properly adjusted according to specific geological conditions.

(4) From a technical perspective, each WCCM method has its own limitations and none of them can be applied individually to the entire mining area. In terms of applicability, backfill mining and height-restricted mining have the widest applicability, followed by partial backfill mining, and narrow strip mining has the least. Considering the low recovery rate associated with height-restricted mining, priority should be given to backfill mining when conditions apply. The applicability zones of each method are compatible with each other rather than mutually exclusive, there are overlaps between the zones suitable for different methods. The optimal WCCM method for the overlap areas should be determined with due consideration of personnel, equipment, and techniques available. Multiple WCCM methods can be used together in order to maximize economic benefits while protecting water resources. For instance, after narrow strip mining, backfill mining can be used to recover the coal pillars left. The combination of these two methods can control the development of water conductive fractures while improving coal recovery rate.

9. Conclusions

(1) A four-level AHP model was constructed to evaluate the influencing factors of WCCM and the weight of each influencing factor was obtained. The results show that: the overburden thickness, stratigraphic structure, the aquifer thickness, aquiclude thickness and the effective mining height are the most important factors among all influencing factors. These five most important factors are plotted as "five maps", forming the basis for studying WCCM in the Yu-Shen mining area.

(2) Based on the theory analysis, field measurements and numerical simulation results, the equations were developed to predict the mining-induced heights of caved zone, WCFZ and the thickness of protective zone in the mining area. Thus three contour maps of the height of these "three zones" across the Yu-Shen mining area were obtained.

(3) Based on the temporal and spatial relationship between measured bedrock thickness, soil strata thickness, heights of caved zone and WCFZ induced by mining, and the thickness of protective zone, the criterion for determining the impact of coal mining on groundwater were proposed. According to this criterion, the Yu-Shen mining area was divided into six partitions, including extremely severe impacted zone, severe impacted zone, moderate impacted zone, slight impacted zone, unaffected impacted zone and unconfirmed zone.

(4) If the thickness of soil strata plus the thickness of bedrock is greater than the height of WCFZ plus the height of protective zone, then the WCCM can be realized. Subtracting the necessary thickness of protective zone from the total thickness of soil and bedrock yields the height of WCFZ. The maximum allowable mining height (MAMH) for successful WCCM was determined by back calculation, and a contour map of MAMH was drawn for the Yu-Shen mining area.

(5) The case studies proved that applicability partition of different WCCM methods, including height-restricted mining, (partial) backfilling and narrow strip mining, was a reliable tool to guide future mining in the coal area.

(6) Based on this research, it can be concluded that a "five maps, three zones and two zoning plans" approach can be used in other coal mining regions in Northwest China, to optimize mining whilst limiting water disruption above the mining area.

Author Contributions: Y.X. conceived the research and wrote the original draft. L.M. revised and reviewed the manuscript. Y.Y. collected the borehole data. All authors have read and agreed to the published version of the manuscript.

Funding: The Fundamental Research Funds for the Central Universities (2017XKQY096); the Priority Academic Program Development of Jiangsu Higher Education Institutions.

Acknowledgments: This work was supported by the Fundamental Research Funds for the Central Universities (2017XKQY096), and the Priority Academic Program Development of Jiangsu Higher Education Institutions.

Conflicts of Interest: The authors declare that they have no competing financial interests in connection with the work submitted.

References

1. Gu, D.Z. Theory framework and technological system of coal mine underground reservoir. *J. China Coal Soc.* **2015**, *40*, 239–246.
2. Ma, L.; Jin, Z.; Liang, J.; Sun, H.; Zhang, N.; Li, P. Simulation of water resource loss in short-distance coal seams disturbed by repeated mining. *Environ. Earth Sci.* **2015**, *74*, 5653–5662. [CrossRef]
3. Meng, Z.; Shi, X.; Li, G. Deformation, failure and permeability of coal-bearing strata during longwall mining. *Eng. Geol.* **2016**, *208*, 69–80. [CrossRef]
4. Wang, L.; Mu, Y.; Zhang, Q.; Zhang, X. Groundwater use by plants in a semi-arid coal-mining area at the Mu Us Desert frontier. *Environ. Earth Sci.* **2012**, *69*, 1015–1024. [CrossRef]
5. Yin, L.; Hou, G.; Dou, Y.; Tao, Z.; Li, Y. Hydrogeochemical and isotopic study of groundwater in the Habor Lake Basin of the Ordos Plateau, NW China. *Environ. Earth Sci.* **2009**, *64*, 1575–1584. [CrossRef]
6. Adhikary, D.; Guo, H. Modelling of longwall mining-induced strata permeability change. *Rock Mech. Rock Eng.* **2014**, *48*, 345–359. [CrossRef]
7. Booth, C. Groundwater as an environmental constraint of longwall coal mining. *Environ. Earth Sci.* **2006**, *49*, 796–803. [CrossRef]
8. Booth, C.J. Confined-unconfined changes above longwall coal mining due to increases in fracture porosity. *Environ. Eng. Geosci.* **2007**, *13*, 355–367. [CrossRef]
9. Fan, G.; Zhou, L. Mining-induced variation in water levels in unconsolidated aquifers and mechanisms of water preservation in mines. *Min. Sci. Technol. (China)* **2010**, *20*, 814–819. [CrossRef]
10. Gale, W.J. *Aquifer Inflow Prediction above Longwall Panels, ACARP End of Grant Report*; ACARP: Brisbane, Australia, 2008; C13013; p. 96.
11. Li, Y. Groundwater system for the periods of pre- and post-longwall mining over thin overburden. *Int. J. Min. Reclam. Environ.* **2015**, *30*, 295–311. [CrossRef]
12. Ma, L.Q.; Sun, H.; Wang, F.; Li, J.M.; Jin, Z.Y.; Zhang, W. Analysis of the ground water level change of aquifer-protective mining in longwall coalface for shallow seam. *J. Min. Saf. Eng.* **2014**, *31*, 232–235.
13. Majdi, A.; Hassani, F.P.; Nasiri, M.Y. Prediction of the height of destressed zone above the mined panel roof in longwall coal mining. *Int. J. Coal Geol.* **2012**, *98*, 62–72. [CrossRef]
14. Fan, L.M. Discussing on coal mining under water-containing condition. *Coal Geol. Explor.* **2005**, *33*, 50–53.
15. Fan, L.M. Scientific connotation of water-preserved mining. *J. China Coal. Soc.* **2017**, *42*, 27–35.
16. Qian, M.G.; Xu, J.L.; Wang, J.C. Further on the sustainable mining of coal. *J. China Coal Soc.* **2018**, *43*, 1–13.
17. Zhang, S.; Fan, G.; Liu, Y.; Ma, L. Field trials of aquifer protection in longwall mining of shallow coal seams in China. *Int. J. Rock Mech. Min. Sci.* **2010**, *47*, 908–914. [CrossRef]
18. Karaman, A.; Akhiev, S.S.; Carpenter, P.J. A new method of analysis of water-level response to a moving boundary of a longwall mine. *Water Resour. Res.* **1999**, *35*, 1001–1010. [CrossRef]
19. Shultz, R. Ground-water hydrology of Marshall County, West Virginia, with emphasis on the effects of longwall coal mining. *Water Resour. Investig. Rep.* **1988**. [CrossRef]
20. Booth, C.J.; Spande, E.D.; Pattee, C.T.; Miller, J.D.; Bertsch, L.P. Positive and negative impacts of longwall mine subsidence on a sandstone aquifer. *Environ. Earth Sci.* **1998**, *34*, 223–233. [CrossRef]

21. Hill, J.G.; Price, D.R. The impact of deep mining on an overlying aquifer in western pennsylvania. *Ground Water Monit. Remediat.* **1983**, *3*, 138–143. [CrossRef]

22. Robertson, J. Challenges in sustainably managing groundwater in the Australian Great Artesian Basin: Lessons from current and historic legislative regimes. *Hydrogeol. J.* **2019**, *28*, 343–360. [CrossRef]

23. Howladar, M.F. Coal mining impacts on water environs around the Barapukuria coal mining area, Dinajpur, Bangladesh. *Environ. Earth Sci.* **2012**, *70*, 215–226. [CrossRef]

24. Gandhe, A.; Venkateswarlu, V.; Gupta, R.N. Extraction of coal under a surface water body—A strata control investigation. *Rock Mech. Rock Eng.* **2005**, *38*, 399–410. [CrossRef]

25. Fan, L.M.; Xiang, M.X.; Peng, J.; Li, C.; Li, Y.H.; Wu, B.Y.; Bian, H.Y.; Gao, S.; Qiao, X.Y. Groundwater response to intensive mining in ecologically fragile area. *J. China Coal Soc.* **2016**, *41*, 2672–2678.

26. Wang, S.M.; Huang, Q.X.; Fan, L.M.; Yang, Z.Y.; Shen, T. Study on overburden aquiclude and water protection mining regionazation in the ecological fragile mining area. *J. China Coal Soc.* **2010**, *35*, 7–14.

27. Miao, X.X.; Chen, R.H.; Bai, H.B. Fundamental concepts and mechanical analysis of water resisting key strata in water preserved mining. *J. China Coal Soc.* **2017**, *32*, 561–564.

28. Huang, Q.X. Research on roof control of water conservation mining in shallow seam. *J. China Coal Soc.* **2017**, *42*, 50–55.

29. Zhang, D.S.; Li, W.P.; Lai, X.P.; Fan, G.W.; Liu, W.Q. Development on basic theory of water protection during coal mining in northwest of China. *J. China Coal Soc.* **2017**, *42*, 36–43.

30. Ma, L.Q.; Zhang, D.S.; Dong, Z.Z. Evolution mechanism and process of aquiclude fissures. *J. Min. Saf. Eng.* **2011**, *28*, 340–344.

31. Ma, L.Q.; Cao, X.Q.; Liu, Q.; Zhou, T. Simulation study on water-preserved mining in multi-excavation disturbed zone in close-distance seams. *Environ. Eng. Manag. J.* **2013**, *12*, 1849–1853. [CrossRef]

32. Ma, L.Q.; Du, X.; Wang, F.; Liang, J.M. Water-preserved Mining Technology for Shallow Buried Coal Seam in Ecologically-vulnerable Coal field: A case study in the Shendong Coal field of China. *Disaster Adv.* **2013**, *6*, 268–278.

33. Wang, A.; Ma, L.; Wang, Z.; Zhang, D.; Li, K.; Zhang, Y.; Yi, X. Soil and water conservation in mining area based on ground surface subsidence control: Development of a high-water swelling material and its application in backfilling mining. *Environ. Earth Sci.* **2016**, *75*, 779. [CrossRef]

34. Li, M.; Zhang, J.; Huang, P.; Gao, R. Mass ratio design based on compaction properties of backfill materials. *J. Cent. South. Univ.* **2016**, *23*, 2669–2675. [CrossRef]

35. Wu, Q. Progress, problems and prospects of prevention and control technology of mine water and reutilization in China. *J. China Coal Soc.* **2014**, *39*, 795–805.

36. Zhang, D.S.; Liu, H.L.; Fan, G.W.; Wang, X.F. Connotation and prospection on scientific mining of large Xinjiang coal base. *J. Min. Saf. Eng.* **2015**, *32*, 1–6.

37. Zhang, S.; Fan, G.; Ma, L.; Wang, X. Aquifer protection during longwall mining of shallow coal seams: A case study in the Shendong Coalfield of China. *Int. J. Coal Geol.* **2011**, *86*, 190–196. [CrossRef]

38. Li, W.P.; Ye, G.J.; Zhang, L.; Duan, Z.H.; Zhai, L.J. Study on the engineering geological conditions of protected water resources during coal mining action in Yu-Shen Mine Area in the North Shanxi Province. *J. China Coal Soc.* **2000**, *25*, 449–454.

39. Du, J.P.; Meng, X.R. *Mining Science*; China University of Mining and Technology Press: Xuzhou, China, 2014.

40. Peng, S.Y. *Coal Mine Ground Control*; China University of Mining and Technology Press: Xuzhou, China, 2013.

41. Sui, W.; Hang, Y.; Ma, L.; Wu, Z.; Zhou, Y.; Long, G.; Wei, L. Interactions of overburden failure zones due to multiple-seam mining using longwall caving. *Bull. Int. Assoc. Eng. Geol.* **2014**, *74*, 1019–1035. [CrossRef]

42. Miao, J.K.; Jiang, Z.Q. Research on the height of water flowing fractured zone in the fully mechanized mining face by drilling exploration. *Shaanxi Coal* **2014**, *33*, 33–36.

43. Wei, J.C.; Wu, F.Z.; Xie, D.L.; Yin, H.Y.; Guo, J.B.; Xiao, L.L. Development characteristic of water flowing fractured zone under semi-cemented medium-low strength country rock. *J. China Coal Soc.* **2016**, *41*, 974–983.

44. Yin, S.X.; Xu, B.; Xu, H.; Xia, X.X. The research on the height of the "Two zones" in Zhuozishan coalfield. *Disaster Adv.* **2013**, *6*, 78–84.

45. Gao, B.B.; Liu, Y.P.; Pan, J.Y.; Yuan, T. Detection and analysis of height of water flowing fractured zone in underwater mining. *Chin. J. Rock Mech. Eng.* **2014**, *33*, 3384–3390.

46. Huang, H.F.; Yan, Z.G.; Yao, B.H.; Xu, H.J. Research on the process of fracture development in overlying rocks under coal seams group mining in Wanli mining area. *J. Min. Saf. Eng.* **2012**, *29*, 619–624.

47. Wang, F.; Tu, S.; Zhang, C.; Zhang, Y.; Bai, Q. Evolution mechanism of water-flowing zones and control technology for longwall mining in shallow coal seams beneath gully topography. *Environ. Earth Sci.* **2016**, *75*, 1309. [CrossRef]

48. Xu, Y.C. Design methods of the effective water-resisting thickness for the protective seam of the water barrier in fully-caving mechanized coalmining. *J. China Coal Soc.* **2005**, *30*, 305–308.

49. Fan, L.M.; Ma, X.D.; Ji, R.J. The progress of research and engineering practice of water-preserved coal mining in western eco-environment frangible area. *J. China Coal Soc.* **2015**, *40*, 1711–1717.

50. Qian, M.G.; Miao, X.X.; Xu, J.L. Green mining of coal resources harmonizing with environment. *J. China Coal Soc.* **2007**, *32*, 1–7.

51. Shi, B.Q. Research on water-preserved-mining in shallow seam covered with rock soil and sand in Northern Shaanxi. *J. Min. Saf. Eng.* **2011**, *28*, 548–552.

52. Lv, W.H. The application of the backfill mining in Yu-yang coal area. *Sci. Technol. Innov. Her.* **2013**, *33*, 48–50.

53. Wang, W.X.; Sui, W.; Faybishenko, B.; Stringfellow, W. Permeability variations within mining-induced fractured rock mass and its influence on groundwater inrush. *Environ. Earth Sci.* **2016**, *75*, 326. [CrossRef]

54. Chi, M.; Zhang, D.; Liu, H.; Wang, H.; Zhou, Y.; Zhang, S.; Yu, W.; Luang, S.; Zhao, Q. Simulation analysis of water resource damage feature and development degree of mining-induced fracture at ecologically fragile mining area. *Environ. Earth Sci.* **2019**, *78*, 88. [CrossRef]

55. Lu, Y.; Wang, L. Numerical simulation of mining-induced fracture evolution and water flow in coal seam floor above a confined aquifer. *Comput. Geotech.* **2015**, *67*, 157–171. [CrossRef]

56. Hollá, L.; Buizen, M. The ground movement, strata fracturing and changes in permeability due to deep longwall mining. *Int. J. Rock Mech. Min. Sci. Géoméch. Abstr.* **1991**, *28*, 207–217. [CrossRef]

Article

Acoustic Emission Characteristics of Coal Samples under Different Stress Paths Corresponding to Different Mining Layouts

Yiming Yang [1], Ting Ai [2,*], Zetian Zhang [1], Ru Zhang [1], Li Ren [2], Jing Xie [1] and Zhaopeng Zhang [2]

[1]   State Key Laboratory of Hydraulics and Mountain River Engineering, College of Hydraulic and Hydropower Engineering, Sichuan University, Chengdu 610065, China; 2018223065136@stu.scu.edu.cn (Y.Y.); zhangzetian@scu.edu.cn (Z.Z.); zhangru@scu.edu.cn (R.Z.); xiejing200655@163.com (J.X.)
[2]   MOE Key Laboratory of Deep Earth Science and Engineering, School of Architecture and Environment, Sichuan University, Chengdu 610065, China; renli-scu@hotmail.com (L.R.); zhangzp@scu.edu.cn (Z.Z.)
*   Correspondence: aiting@scu.edu.cn; Tel.: +86-1780-105-1840

Received: 11 April 2020; Accepted: 20 June 2020; Published: 26 June 2020

Abstract: Research on the mining-induced mechanical behavior and microcrack evolution of deep-mined coal has become increasingly important with the sharp increase in mining depth. For rock units in front of the working face, the microcrack evolution characteristics, structural characteristics, and stress state correspond well to mining layouts and depths under deep mining. The acoustic emission (AE) characteristics of typical coal under deep mining were obtained by conducting laboratory experiments to simulate mining-induced behavior and utilizing AE techniques to capture the variation in AE temporal and spatial parameters in real time, which provide an important basis for studying the rupture mechanisms and mechanical behavior of deep-mined coal. The findings were as follows: (1) AE activity under deep mining was characterized by three stages, corresponding to crack initiation, crack stable propagation, and crack unstable propagation. As the three stages proceeded, the AE counting rate and AE energy rate presented stronger clustering characteristics, and the cumulative AE counting and cumulative AE energy exhibited a sharp increase by an order of magnitude. (2) The crack initiation and the main stages of crack propagation were determined by characteristic points of variation curves in the AE parameters over time. In the main crack propagation stage, the number of cumulative AE events and the cumulative AE counts were similar among the three mining conditions, while coal samples under coal pillar mining released the largest amount of AE energy. The amount of accumulated AE energy released by coal samples increased by one order of magnitude according to the sequence of protective coal-seam mining, top-coal caving mining, and nonpillar mining. (3) Fractal technology was applied to quantitatively analyze the AE spatial evolution process, showing that the fractal dimension of the AE location decreased as the peak stress increased, corresponding to protective seam mining, caving-coal mining, and nonpillar mining. The above results showed that the deformation and fracture characteristics of coal under deep mining followed a general law, but were affected by different mining conditions. The crack initiation and main rupture activity of coal occurred earlier under the conditions of protective seam mining, top-coal caving mining, and nonpillar mining, successively. Moreover, nonpillar mining induced the strongest and highest degree of unstable rupture of the coal body in front of the working face.

Keywords: acoustic emission (AE); coal; deep mining; mining layouts

---

1. Introduction

Worldwide, recoverable coal deposits will last just over 150 years at the current consumption rate [1]; coal energy is playing an increasingly important role in supplying the primary energy of developing countries such as China [2]. Coal is the dominant energy source in China, accounting for 76% and 66% of the primary energy production and consumption structure, respectively. In China, coal mining activity has deepened at a rate of 0 to 25 m annually, with the mining depths of major mines in the central and eastern regions of China reaching 800 to 1000 m [3–5]. As the mining depth increases, the stress of the coal mass increases accordingly, which leads to phenomena such as a high ground stress, high ground temperature, and high osmotic pressure, thereby inducing disasters such as rock bursts, water inrushes, and coal and gas outbursts. The rheology of the surrounding rock makes the roadway difficult to support and causes the associated maintenance costs to increase dramatically [6]. To meet the worldwide demand for coal resources and ensure the safe and reliable development of deep coal resources, it is particularly important to conduct research on issues related to deep coal mining. In the past five years, many researchers have investigated various aspects of deep coal mining. Yang et al. [7] carried out numerical simulations to study the mechanism for the large deformation of a deep soft rock roadway. Sun et al. [8] studied the relationship between deep mining and surface subsidence, and analyzed coal pillar safety for strip mining in deep mines. Kong et al. [3], Wang et al. [4], Liu et al. [9] and Zhou et al. [10] carried out hydraulic fracturing in a high gassy coal seam and reduced the coal seam gas content through surface shafts and underground deep holes to eliminate the risk of coal and gas outbursts during the initial mining.

In situ coal in stress equilibrium exhibits the phenomena of deformation, failure, shock and instability under mining operations. Therefore, the mechanical behavior of deep coal is closely related to mining layouts and mining disturbances [11]. In the actual mining process, the coal mass in front of the working face experiences a complete mining-induced mechanical unloading process, in which the in situ stress and axial stress increase to failure while the confining pressure decreases [12]. Recently, researchers conducted a series of studies on the mining-induced mechanical behavior of coal under different mining conditions. Xie et al. [11] carried out mining-induced mechanical tests, and showed that, under different mining layouts, the peak stress of coal and its corresponding axial strain and circumferential strain decrease, while the volumetric strain increases in the order of nonpillar mining, top-coal caving mining and protective coal-seam mining. Zhang et al. [13] studied the structural and connectivity characteristics of coal sample ruptures under different mining layouts. Zhang et al. [14] conducted laboratory tests and in situ observations from which they derived a theoretical expression of the mining-induced permeability change ratio, observing that the stress evolution process under different mining layouts affects the mechanical behavior and permeability change rate of the coal mass. Zhang et al. [15] studied the fractal characteristics of fractures under different mining layouts, and found that a relatively large volume of fractures can be generated in nonpillar mining, resulting in a large number of fractures in front of the working face. Overall, based on the consideration of the mining-induced mechanical behavior of coal, it was found that there are notable differences, in terms of variations in strength characteristics, damage and permeability, among coal under different mining layouts. Additionally, the yield criterion and failure mechanism of coal, considering the mining process, were established. However, the existing research outcomes were not derived based on deep conditions. The influence of different mining layouts on the mechanical properties and failure mechanism of coal mass under the deep environment of high ground stress, high ground temperature and high osmotic pressure is still unclear. To address this problem, it is necessary to simultaneously study the deep environment and the mining conditions associated with deep coal mining.

Research on the mining-induced mechanical behavior of deep coal has usually focused on crack evolution and the failure characteristics of coal. Microcrack propagation and the failure process of rock are closely related to its acoustic emission (AE) behavior [16–19]. In recent years, three-dimensional (3D) AE localization technology has been widely used in laboratory mechanical testing, and has proven to be an effective method for studying crack propagation and failure characteristics [20,21]. Su et al. [22] studied the AE and temperature variations of coal, and demonstrated that AE is the result of thermal stress-induced crack formation and propagation. Agioutantis et al. [23] used AE to predict rock damage, and showed that the local rock damage can be determined based on the occurrence of AE events in the critical stress region. Coal exhibits poor homogeneity and complex mechanical properties compared to other rocks, leading to some deficiencies in the current implementation of 3D AE localization technology. Nevertheless, the technology has gradually matured and is now generally stable. Many studies involving laboratory mechanical tests of coal and simultaneous 3D AE localization have been published [17,24–26]. Because the mining-induced mechanical behavior of deep coal is related mainly to unloading, it is, to some extent, still experimentally difficult to simultaneously use AE localization technology while conducting a laboratory unloading experiment for mining simulation. As a result, only a limited amount of literature exists in this area [17,24,27–29].

It is generally believed that deep mining corresponds to a coal-seam burial depth greater than 800 m [30,31]. In this study, a series of laboratory experiments were carried out to simulate the mining process of coal at a burial depth of 1000 m, and real-time monitoring of this process was conducted for simultaneous AE localization. The time series activities, spatial localization evolution and fractal characteristics of AE under different deep mining conditions were studied to deepen the understanding of the mining-induced mechanical behavior and rupture process of deep coal. In the study, three deep mining conditions, i.e., protective coal-seam mining [32], top-coal caving mining [33], and non-pillar mining [34], were considered.

2. Experimental Area

2.1. Sample Specification and AE Sensor

Samples were taken from working face 8105 of the Tashan Mine, operated by Datong Coal Industry Group Co., Ltd. The coal seam has an average thickness of 18.17 m. The coal was processed into cylindrical samples with a height of 100 mm and a diameter of 50 mm. Coal samples are relatively difficult to process, so there is always a certain amount of error between the actual and ideal samples. The statistical analysis of the 9 coal samples used in the test showed diameters in the range of 49.90 to 50.42 mm, and heights in a range of 98.53 to 100.46 mm. Some of the processed samples are shown in Figure 1.

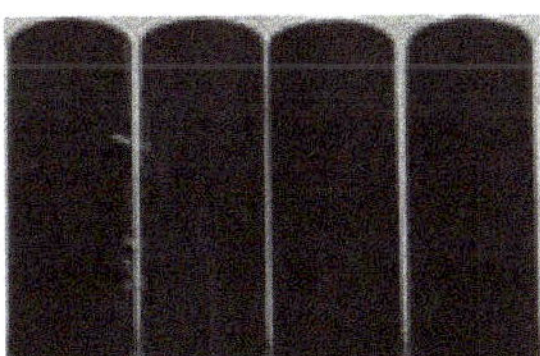

Figure 1. Coal samples.

Mechanical loading was performed using an MTS815 Flex test GT concrete and rock mechanics test system. An AE system (PCI-2, Physical Acoustics Co./PAC) was used for data acquisition. The test setup was exhibited in Figure 2. As shown in Figure 3, eight AE sensors were evenly placed on the outer wall surface of the triaxial pressure chamber (an upper and lower circle of four sensors each). The sensors were coupled with the wall of the pressure chamber by coating their contact area with Vaseline and fixing them using a rubber ring, and were used for real-time monitoring of the time parameters and spatial location of the AE events. The AE sensor model was a Nano30 resonator sensor, with a peak frequency of 300 kHz and a working frequency band of 100 to 500 kHz. Because the sensor was small (see Figure 3) and the chip was at the center of the probe, Vaseline was used as an adhesive to ensure that the chip of the sensor probe fully contacted the measured point, thus ensuring that AE signals could be received. In the test, the wave velocity was 1500 m/s and the threshold was 40 dB.

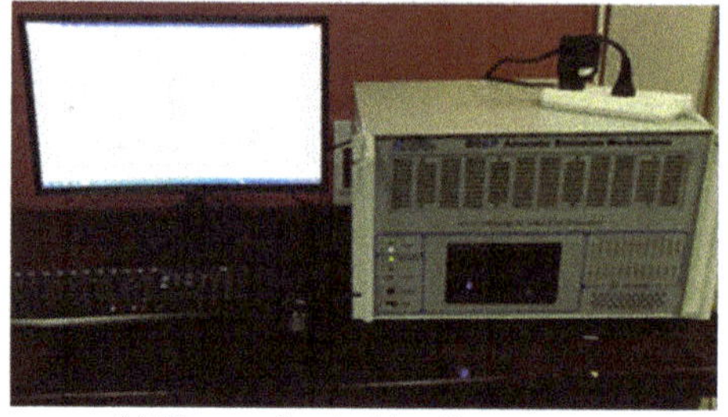 

(**a**) The mechanical loading system   (**b**) AE system

Figure 2. Test setup.

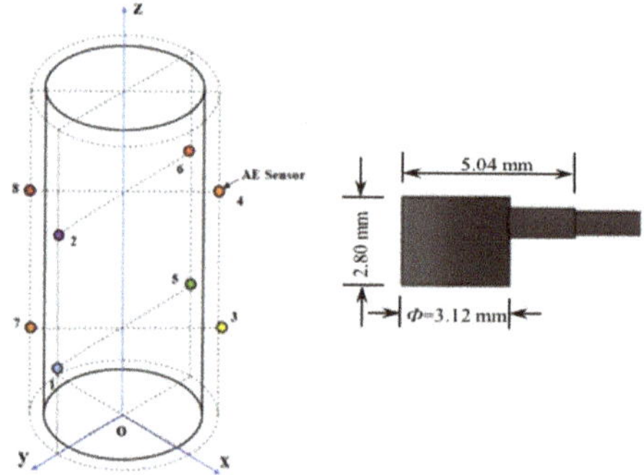

Figure 3. Distributions and dimensions of an AE sensor.

2.2. Test Process

According to various data collected from field tests, Xie et al. [12,35] summarized the variation characteristics of the abutment pressure of the coal mass in front of the working face induced by actual mining activities. The stress concentration coefficient α under different mining layouts differed under the condition of similar roof property and mining depth, which were 2.5–3.0, 2.0–2.5, and 1.5–2.0 respectively, in the order of nonpillar mining, top-coal caving and protective seam mining. Nonpillar mining is a longwall mining process in which there is no mining roadway and no reserves of coal pillars, while protective seam mining is characterized by mining the protective layer in advance, thus disturbing the adjacent coal seam to weaken its risk of coal and gas outburst. In the simulation experiment, it was assumed that α was 3.0, 2.5 and 2.0 for nonpillar mining, top-caving mining and protected seam mining, respectively. Therefore, based on the theory proposed by Xie et al. [12,36], the entire mining-induced mechanical loading (unloading) process can be divided into three stages: the hydrostatic pressure stage (OA), the first unloading stage (AB) and the second unloading stage (BC, BD, or BE), as shown in Figure 4. In Figure 4, γ refers to the average specific gravity of rocks on top of the coal seam and f_c' is the triaxial compressive strength of coal. The simulated mining depth was approximately 1000 m and the hydrostatic confining pressure in OA was defined as 25 MPa; from this, the horizontal and vertical stress in each stage can be calculated.

Figure 5 exhibits the loading (unloading) scheme corresponding to different mining layouts at 1000 m depth. In the first stage, to simulate the influence of initial hydrostatic pressure on coal rock before mining, the hydrostatic confining pressure was applied to an in situ stress state at point A ($\sigma_1 = \sigma_3 = \gamma H = 25$ MPa) at the same confining pressure loading rate of 3 MPa/min under each of the three mining layouts. The second stage was the first unloading stage. With the occurrence of excavation disturbance, the axial stress of coal rock increased and the horizontal stress decreased. All three groups of coal samples were unloaded, starting from point A, at a confining pressure unloading rate of 1

MPa/min; at the same time, all were loaded to point B ($\sigma_1 = 1.5\,\gamma H = 37.5$ MPa, $\sigma_3 = 0.6\,\gamma H = 15$ MPa) at an axial pressure loading rate of 2.25 MPa/min. In other words, the axial stress was increased at a rate of 2.25:1 and the lateral stress was reduced. After point B, the specimen was divided into three loading paths to reflect the stress effect produced by different mining methods. The three mining layouts were associated with the same confining pressure unloading rate of 1 MPa/min and axial loading rates of 2.25 MPa/min (protective seam), 3.5 MPa/min (top-coal caving) and 4.75 MPa min (nonpillar) until sample failure.

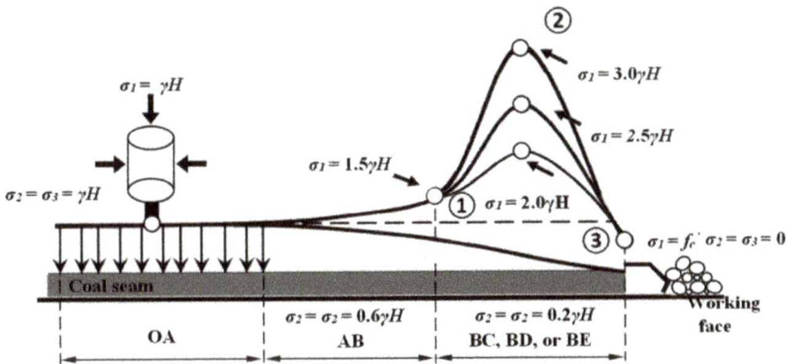

Figure 4. The stress state of coal mass in front of the working faces under three different mining layouts.

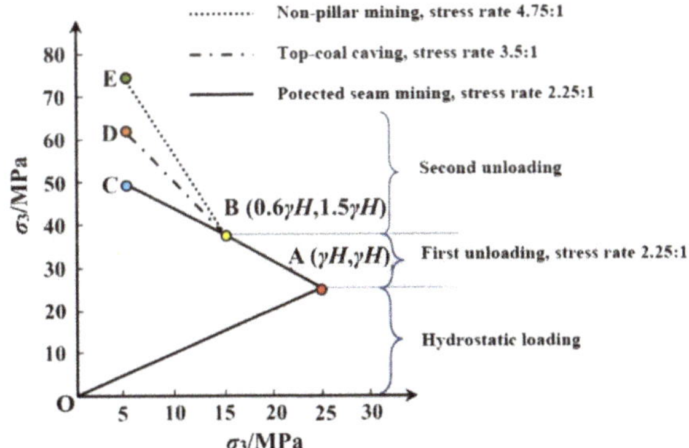

Figure 5. The loading (unloading) scheme under different mining layouts ($\gamma H = 25$ MPa).

3. Results

3.1. Characteristics of AE Time Series Evolution of Coal under Deep Mining Conditions

Of the AE data obtained by simultaneous real-time monitoring, the cumulative ring count and ring count rate reflect the frequency of AE occurrence, and the cumulative energy and energy rate reflect the intensity of AE occurrence. Figure 6 shows the variation in the AE ring-down count rate (a1,b1,c1), cumulative ring-down count (a2,b2,c2), energy rate (a3,b3,c3) and cumulative energy (a4,b4,c4) over time under the simulated deep mining condition at a depth of 1000 m. In Figure 6a–c correspond to protective seam mining, top-coal caving and nonpillar mining, respectively.

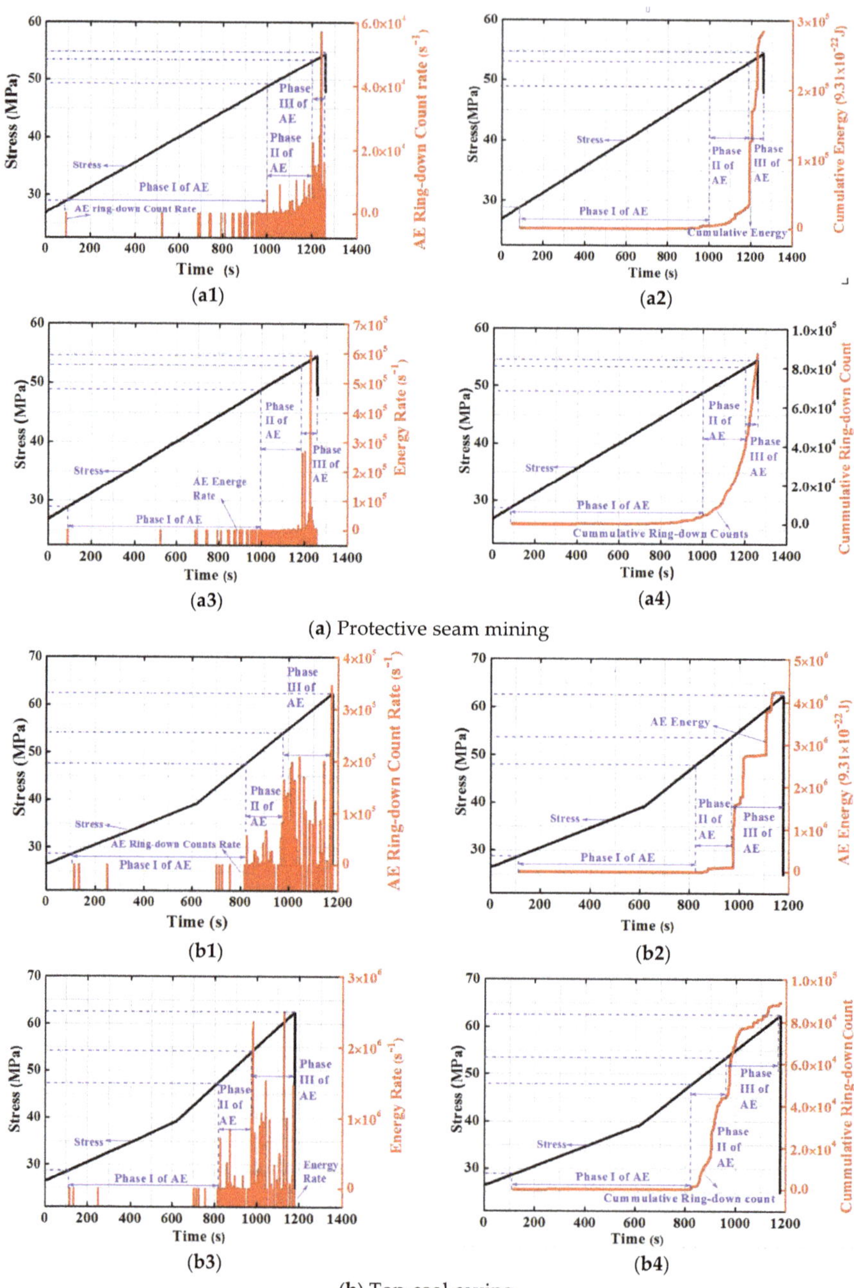

(**a1**)

(**a2**)

(**a3**)

(**a4**)

(**a**) Protective seam mining

(**b1**)

(**b2**)

(**b3**)

(**b4**)

(**b**) Top-coal caving

Figure 6. *Cont.*

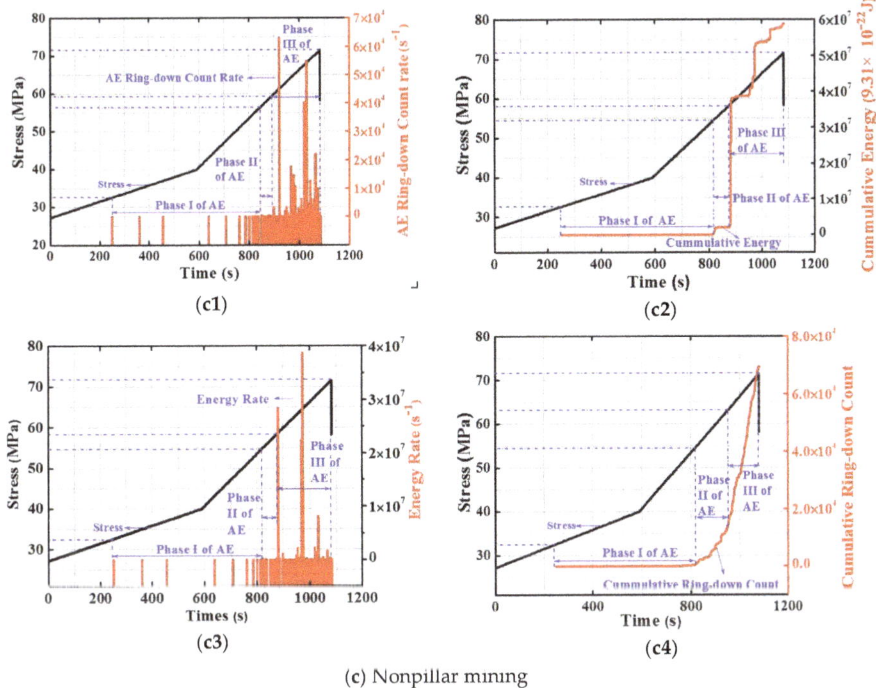

Figure 6. Evolution of the AE ring-down count rate, cumulative energy, energy rate and cumulative ring-down count over time.

For rocks with prefabricated cracks, the crack initiation point and propagation process are the usual targets of research [37,38]. For cylinder specimens without macroscopic cracks, Cai et al. [27] concluded that rock experiences four stages: crack initiation, microcrack formation, crack coalescence and macrocrack formation under compression conditions. Coal has a complex structure with viscoelastic characteristics such as instantaneous deformation, elastic hysteresis and irreversible plastic deformation. Hence, its mechanical behavior is also complex, and cannot be simply described by the five-stage constitutive rock model. Studying the characteristics of the AE time series evolution curve can make the research results of the crack initiation point and propagation process more convincing. Figure 6 shows that the overall AE activity exhibited three characteristic stages over time. During the first stage of AE activity, the overall AE activity was at a low level. In the second stage, the amplitudes of the AE ring-down count rate and energy rate showed a significant increase compared to those in the previous stage on average, and began to exhibit a cluster characteristic with time, but had a small range of fluctuation (see Figure 6 (a1,a3,b1,b3,c1,c3)). The AE cumulative energy and cumulative ring-down count began to increase slowly (see Figure 6 (a2,a4,b2,b4,c2,c4)). In the third stage, the amplitudes of the AE ring-down count rate and the energy ratio changed abruptly compared with those in the previous two stages on average, and exhibited an increasingly strong clustering characteristic with time. In this stage, the cumulative energy curve began to show a relatively high step, increasing by an order of magnitude in a few seconds. It can be concluded that the three variation stages of AE activity correspond to the three processes (i.e., microcrack initiation, stable propagation and unstable propagation) inside the coal.

Statistical analyses were performed on the characteristic points of the three AE activity stages of coal under three mining layouts, as shown in Table 1. The starting point for each of the three AE activity stages corresponding to the three mining layouts (i.e., nonpillar, top-coal caving and protective seam) is associated with an increasing stress level, allowing the occurrence timing of rupture initiation and main rupture activity under the above mining layouts to be ranked in ascending order.

Based on the three stages of AE activities quantitively divided previously, the stress-strain curves can also be divided into three parts. Figure 7a–c shows the relationship between the released AE energy and the axial strain at the three characteristic stages under each of the three mining layouts plotted in the same coordinate system. Combined with Figure 7d, the following can be observed. In the first stage of AE activity, the axial stress and strain under each of the three mining layouts were in an elastic range; the magnitude of released AE energy was on the order of 9.31×10^{-18} to 10^{-20} J. In the second stage, the coal deformation progressed from the elastic stage to the elastoplastic deformation stage; the magnitude of released AE energy increased to the order of 9.3×10^{-15} to 10^{-17} J. In the third stage of AE activity, the stress-strain relationship curve showed an increase in curvature and exhibited a stress plateau, indicating plastic deformation of coal in this stage; the released AE energy was in the range of 10^{-15} to 10^{-17} J. Therefore, it can be seen that, for each of the three mining layouts, the AE energy was mainly released during the second and third stages of AE activity. Thus, these two stages can be considered the main stages of coal rupture.

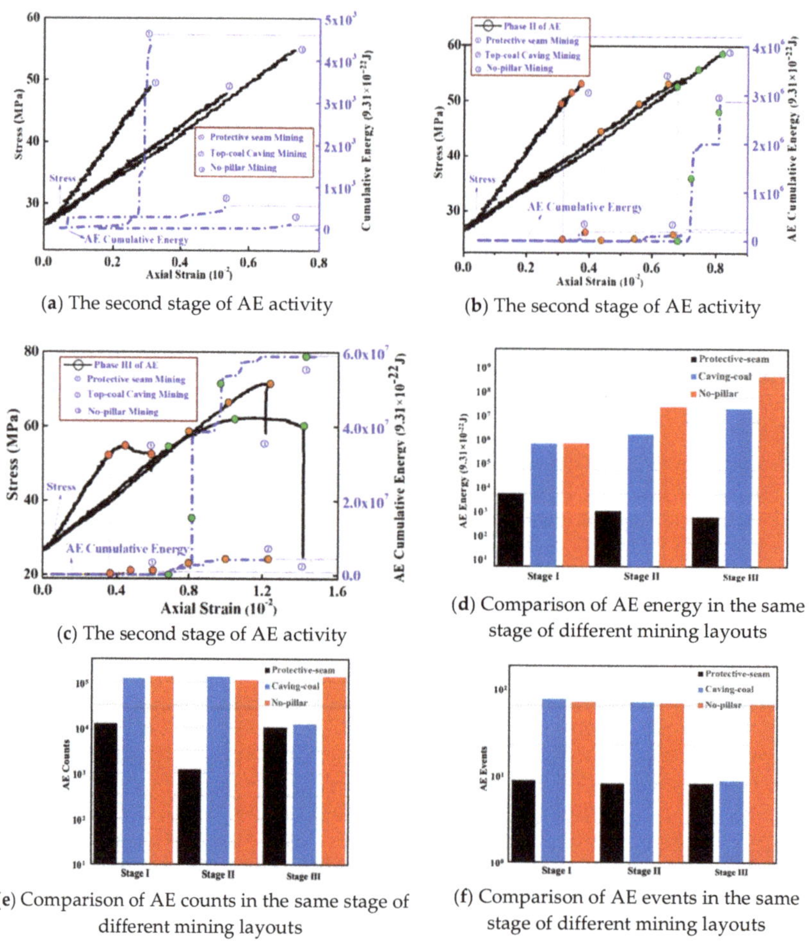

(**a**) The second stage of AE activity

(**b**) The second stage of AE activity

(**c**) The second stage of AE activity

(**d**) Comparison of AE energy in the same stage of different mining layouts

(**e**) Comparison of AE counts in the same stage of different mining layouts

(**f**) Comparison of AE events in the same stage of different mining layouts

Figure 7. AE events, ring-down count, and energy under different mining layouts.

Table 1. Statistics of stress and time for typical stages of AE activity under different mining layouts.

Mining Layouts	Phase I Starting Point		Phase II Starting Point		Phase III Starting Point	
	Stress/%	Time/s	Stress/%	Time/s	Stress/%	Time/s
Protective-seam	53	92.9	89.6	1002.7	97.8	1205.3
Top-coal caving	46.1	117.9	76.7	827.7	86.7	976.1
Nonpillar	45.6	250.2	76.8	823.7	81.9	880.7

3.2. Spatial Evolution and Fractal Characteristics of AE of Coal under Deep Mining Conditions

According to the defined standard for the three-stage evolution of the AE time series, Figure 8 shows the spatial relationship between the AE event and the strain under different mining layouts. In the first stage, AE events were distributed inside the sample, as shown in Figure 8a. At the end of the second stage, the coal material began to yield, and a main AE cluster area formed, as seen in Figure 8b,e. In the subsequent evolution process, the generated AE was concentrated in the main cluster area, as seen in Figure 8c,f, corresponding to the main coalescing area of coal cracks, as shown in Figure 8d. After the completion of the entire experiment, the cumulative morphological distribution of AE spatial localization had a consistent corresponding relationship with the macrocracks in the ruptured coal mass experiencing the complete mining process. The coal failure mode under each of the three mining layouts can be preliminarily determined from Figure 8d as follows: The sample corresponding to the protective coal-seam mining failed under compression, and the coal corresponding to top-coal caving mining and nonpillar mining failed under tension.

Fractal geometry provides a quantitative method for testing spatial morphology. Hirata et al. and Zhang et al. [39,40] used AE localization tests to verify that the spatial distribution of AE events has fractal characteristics. The evolution process of the AE spatial distribution under the influence of coal mining can be viewed as a transition from a disorderly to an orderly state, and the fractal dimension provides us with an accurate order parameter describing the AE activity during such a transition.

The general formula for solving fractal dimension D is

$$\log N = \log a - D \log \delta, \tag{1}$$

Boxes of different sizes δ were used to cover the study objects, and the total number of boxes covering the study objects was denoted as N. A set of (δ, N) data obtained in the covering process is presented in a log-log plot, and the slope is the fractal dimension D.

The cylinder covering method [41] was used for the fractal measurement of AE spatial localization. The mass center of the cylindrical sample was used as a base point, and a small cylinder (with a radius of r and a height of h) proportional to the sample with respect to the height-diameter ratio was taken. By simultaneously increasing r and h, the number of AE localization points covered by each cylinder was counted, and then Equation (1) was used to obtain the cluster dimension of AE spatial location points.

Figure 9 shows the relationship between the AE spatial fractal dimension and the peak stress at the coal peak stress location plotted in the same coordinate system. As the peak stress increased, the AE spatial fractal dimension decreased successively under the protective seam, top-coal caving and nonpillar mining layouts.

In a deep mining process, under the influence of confining pressure unloading (Figure 9), the spatial microcracks were clustered into two-dimensional (2D) variants, resulting in a relatively small fractal dimension corresponding to the peak stress. At the peak stress location, nonpillar mining corresponds to the lowest fractal dimension, which is closest to the 2D stress state, indicating that the nonpillar mining layout resulted in a higher degree of unstable coal mass rupture in front of the working face. The higher degree of coal mass rupture indicates a lower energy accumulation, and implies that the possibility of coal burst risk tends to be low, which coincides with views, based upon practical applications, that nonpillar mining is a relatively safe mining scheme to reduce rock burst likelihood. In nonpillar mining, the working face outside the seam line of the pillar is not reserved, and the roof strata will collapse to the goaf, generating roadside support under the influence of the gangue retaining system behind the support. That process also releases the accumulative energy in the surrounding rock, thereby reducing the risk.

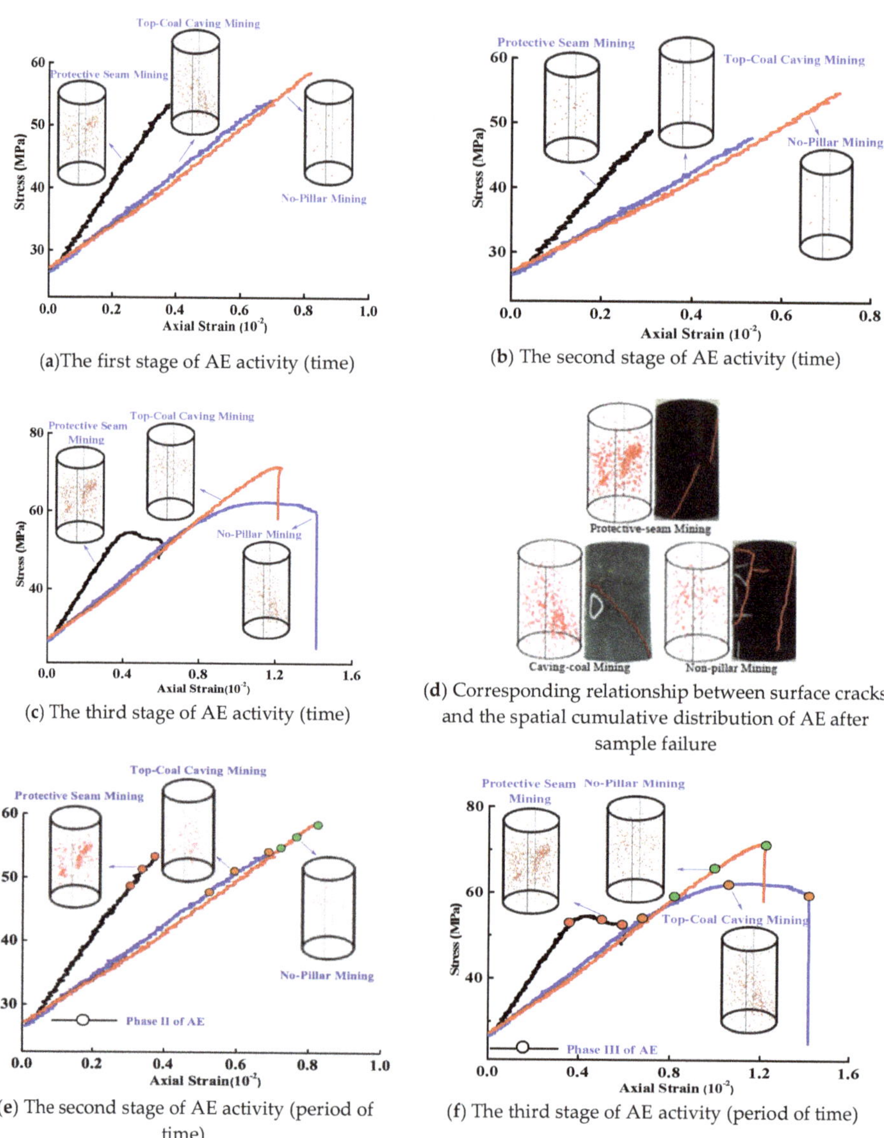

Figure 8. Spatial distribution of AE under different mining layouts.

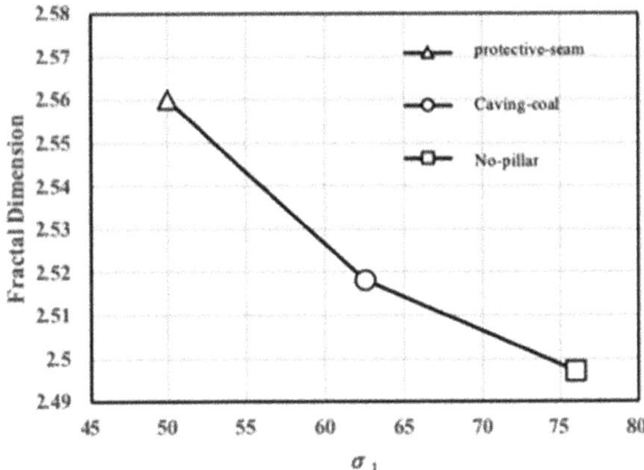

Figure 9. Relationship between AE spatial fractal dimension and stress under different mining conditions.

4. Conclusions

Based on three typical mining layouts, i.e., protective coal-seam, top-coal caving and nonpillar mining, the coal mining process at a depth of 1000 m was simulated through mechanical tests on a laboratory scale, and simultaneous AE localization of this process was carried out to study the AE characteristics under deep coal mining conditions. Based upon this, the following conclusions were drawn:

(1) Under deep mining conditions, the time series evolution of AE activity exhibits three-stage characteristics, which correspond to the processes of initiation, stable propagation and unstable propagation of cracks in the coal.

(2) Under deep mining conditions, the occurrence timing of the main ruptures of the three-stage coal activities may be ranked in descending order, i.e., protective coal seam, top-coal caving and nonpillar mining layouts. The number of cumulative AE events and the number of AE ring-down counts of the three mining layouts were on the same order of magnitude. The accumulated AE energy released by the coal mass in front of the working face of the protective coal seam was one order of magnitude lower than that of caving coal; similarly, the accumulated AE energy of caving coal was lower by one order of magnitude than that of nonpillar mining. The energy released from a single rupture of the sample under the nonpillar layout was higher than that under the other two mining layouts.

(3) Under deep mining conditions, as the peak stress increased, the AE spatial fractal dimension decreased successively under the protective seam, top-coal caving and nonpillar mining layouts. The nonpillar mining layout resulted in a higher degree of unstable coal mass rupture in front of the working face.

Author Contributions: Y.Y., T.A. and Z.Z. contributed to conducting the experiments and writing the manuscript. R.Z. and L.R. modified the writing of the manuscript. J.X. and Z.Z. assisted in the analysis of AE data. All authors have read and agreed to the published version of the manuscript.

Funding: This research was funded by Department of Science and Technology of Sichuan Province (CN) (Grant No. 2017TD0007) and the National Natural Science Foundation of China (Grant No. 51622402).

Acknowledgments: The authors thank anonymous colleagues for their kind efforts and valuable comments, which have improved this work.

Conflicts of Interest: The authors declare no conflict of interest.

References

1. Nicholls, T. *How the Energy Industry Works: An Insider's Guide*; Silverstone Communications Ltd.: London, UK, 2010; pp. 111–112.
2. Han, Y.; Yin, F.; Li, X. A review on water in low rank coals: The existence, interaction with coal structure and effects on coal utilization. *Fuel Process. Technol.* **2013**, *106*, 9–20.
3. Kong, S.; Cheng, Y.; Ren, T.; Liu, H. A sequential approach to control gas for the extraction of multi-gassy coal seams from traditional gas well drainage to mining-induced stress relief. *Appl. Energy* **2014**, *131*, 67–78. [CrossRef]
4. Wang, L.; Cheng, Y.P.; An, F.H.; Zhou, H.X.; Kong, S.L.; Wang, W. Characteristics of gas disaster in the huaibei coalfield and its control and development technologies. *Nat. Hazards* **2014**, *71*, 85–107. [CrossRef]
5. Liang, Y. Strategic thinking of simultaneous exploitation of coal and gas in deep mining. *J. China Coal Soc.* **2016**, *41*, 1–6.
6. Lobanova, T.V. Geomechanical state of the rock mass at the Tashtagol mine in the course of nucleation and manifestation of rock bursts. *J. Min. Sci.* **2008**, *44*, 146–154. [CrossRef]
7. Yang, S.Q.; Chen, M.; Jing, H.W.; Chen, K.F.; Meng, B.A. Case study on large deformation failure mechanism of deep soft rock roadway in Xin'an coal mine, China. *Eng. Geol.* **2016**, *217*, 89–101. [CrossRef]
8. Sun, W.; Zhang, Q.; Luan, Y.; Zhang, X.P. A study of surface subsidence and coal pillar safety for strip mining in a deep mine. *Environ. Earth Sci.* **2018**, *77*, 627. [CrossRef]
9. Liu, T.; Lin, B.; Wei, Y.; Tong, L.; Cheng, Z. An integrated technology for gas control and green mining in deep mines based on ultra-thin seam mining. *Environ. Earth Sci.* **2017**, *76*, 243. [CrossRef]
10. Zhou, A.; Kai, W.; Lei, L.; Wang, C.A. Roadway driving technique for preventing coal and gas outbursts in deep coal mines. *Environ. Earth Sci.* **2017**, *76*, 236. [CrossRef]
11. Xie, H.P.; Zhang, Z.; Gao, F.; Zhang, R.; Gao, M.Z.; Liu, J.F. Stress-fracture-seepage Field Behavior of Coal under Different Mining Layouts. *J. China Coal Soc.* **2016**, *41*, 2405–2417.
12. Xie, H.P.; Zhao, X.; Liu, J.; Zhang, R.; Xue, D. Influence of different mining layouts on the mechanical properties of coal. *Int. J. Min. Sci.* **2012**, *22*, 749–755. [CrossRef]
13. Zhang, R.; Ai, T.; Li, H.; Zhang, Z.; Liu, J. 3D reconstruction method and connectivity rules of fracture networks generated under different mining layouts. *Int. J. Min. Sci. Technol.* **2013**, *23*, 863–871. [CrossRef]
14. Zhang, Z.; Ru, Z.; Xie, H.; Gao, M.; Jing, X. Mining-induced coal permeability change under different mining layouts. *Rock Mech. Rock Eng.* **2016**, *49*, 3753–3768. [CrossRef]
15. Zhang, R.; Ai, T.; Zhou, H.W.; Ju, Y.; Zhang, Z.T. Fractal and volume characteristics of 3D mining-induced fractures under typical mining layouts. *Environ. Earth Sci.* **2015**, *73*, 6069–6080. [CrossRef]
16. Kramadibrata, S.; Matsui, K.; Shimada, H. Role of acoustic emission for solving rock engineering problems in indonesian underground mining. *Rock Mech. Rock Eng.* **2011**, *44*, 281–289. [CrossRef]
17. Ai, T.; Ru, Z.; Liu, J.; Li, R. Space–time evolution rules of acoustic emission location of unloaded coal sample at different loading rates. *Int. J. Min. Sci. Technol.* **2012**, *22*, 847–854.
18. Vishal, V.; Ranjith, P.G.; Singh, T.N. An experimental investigation on behaviour of coal under fluid saturation, using acoustic emission. *J. Nat. Gas Sci. Eng.* **2015**, *22*, 428–436. [CrossRef]
19. Wu, S.; Ge, H.; Wang, X.; Meng, F. Shale failure processes and spatial distribution of fractures obtained by ae monitoring. *J. Nat. Gas Sci. Eng.* **2017**, *41*, 82–92. [CrossRef]
20. Cai, M.; Kaiser, P.K.; Tasaka, Y.; Maejima, T.; Morioka, H.; Minami, M. Generalized crack initiation and crack damage stress thresholds of brittle rock masses near underground excavations. *Int. J. Rock Mech. Min. Sci.* **2004**, *41*, 833–847. [CrossRef]
21. Zhou, H.; Yu, Y.; Zhang, Y.; Lei, F.; Zuo, Y.; Xiang, W. Fine test on progressive fracturing process of multi-crack rock samples under uniaxial compression. *Chin. J. Rock Mech. Eng.* **2010**, *29*, 465–470.
22. Su, F.Q.; Itakura, K.I.; Deguchi, G.; Ohga, K. Monitoring of coal fracturing in underground coal gasification by acoustic emission techniques. *Appl. Energy* **2017**, *189*, 142–156. [CrossRef]
23. Agioutantis, Z.; Kaklis, K.; Mavrigiannakis, S.; Verigakis, M.; Vallianatos, F.; Saltas, V. Potential of acoustic emissions from three point bending tests as rock failure precursors. *Int. J. Min. Sci. Technol.* **2016**, *26*, 155–160. [CrossRef]
24. Liu, Q.; Zhang, R.; Gao, M.; Guo, L.I.; Zhang, Z.; Zhang, Z. Acoustic emission characteristics and comprehensive failure precursors of coal under unloading conditions. *J. Sichuan Univ.* **2016**, *48*, 67–74.

25. Shkuratnik, V.L.; Nikolenko, P.V.; Koshelev, A.E. Spectral characteristics of acoustic emission in loaded coal specimens for failure prediction. *J. Min. Sci.* **2017**, *53*, 818–823. [CrossRef]

26. Zhang, Z.; Zhang, R.; Zhang, Z.; Gao, M.; Dai, F. Experimental research on effect of bedding plane on coal acoustic emission under uniaxial compression. *Chin. J. Rock Mech. Eng. Geol.* **2015**, *34*, 770–778.

27. Shkuratnik, V.L.; Filimonov, Y.L.; Kuchurin, S.V. Experimental investigations into acoustic emission in coal samples under uniaxial loading. *J. Min. Sci.* **2004**, *40*, 458–464. [CrossRef]

28. Shkuratnik, V.L.; Filimonov, Y.L.; Kuchurin, S.V. Regularities of acoustic emission in coal samples under triaxial compression. *J. Min. Sci.* **2005**, *41*, 44–52. [CrossRef]

29. Wang, J.; Xie, L.; Xie, H.; Li, R.; Bo, H.; Li, C. Effect of layer orientation on acoustic emission characteristics of anisotropic shale in Brazilian tests. *J. Nat. Gas Sci. Eng. Geol.* **2016**, *36*, 1120–1129. [CrossRef]

30. Xie, H.P.; Gao, F.; Ju, Y.; Gao, M.Z.; Zhang, R.; Gao, Y.N. Quantitative definition and investigation of deep mining. *J. China Coal Soc.* **2015**, *40*, 1–10.

31. Xie, H.P.; Zhou, H.W.; Xue, D.J.; Wang, H.W.; Zhang, R.; Gao, F. Research and consideration on deep mining and limit mining depth of coal. *J. China Coal Soc.* **2012**, *37*, 535–542.

32. Min, T.U.; Miao, X.X.; Colliery, Z. Deformation rule of protected coal seam exploited by using the long-distance-lower protective seam method. *J. Min. Saf. Eng.* **2006**, *23*, 253–257.

33. Hao, Y.; Zhang, J.; Nan, Z.; Sheng, Z.; Dong, X. Shaft failure characteristics and the control effects of backfill body compression ratio at ultra-contiguous coal seams mining. *Environ. Earth Sci.* **2018**, *77*, 458.

34. Wang, J.C. Fully mechanized longwall top coal caving technology in china and discussion on issues of further development. *Coal Sci. Technol.* **2005**, *33*, 14–17.

35. Xie, H.P.; Zhou, H.W.; Liu, J.F.; Gao, F.; Zhang, R.; Xue, D.J.; Zhang, Y. Mining-induced Mechanical Behavior in Coal Seams under Different Mining Layouts. *J. China Coal Soc.* **2011**, *36*, 1067–1074.

36. Ju, Y.; Zhang, Q.; Zheng, J. Experimental study on CH4 permeability and its dependence on interior fracture networks of fractured coal under different excavation stress paths. *Fuel* **2017**, *202*, 483–493. [CrossRef]

37. Luo, Y.; Ren, L.; Xie, L.Z.; Ai, T.; He, B. Fracture behavior investigation of a typical sandstone under mixed-mode I/II loading using the notched deep beam bending method. *Rock Mech. Rock Eng.* **2017**, *50*, 1–19. [CrossRef]

38. Ren, L.; Xie, L.Z.; Xie, H.P.; Ai, T.; He, B. Mixed-mode fracture behavior and related surface topography feature of a typical sandstone. *Rock Mech. Rock Eng.* **2016**, *49*, 1–17. [CrossRef]

39. Hirata, T.; Satoh, T.; Ito, K. Fractal structure of spatial distribution of micro-fracturing in rock. *Geophys. J. Int.* **1987**, *90*, 369–374. [CrossRef]

40. Zhang, R.; Dai, F.; Gao, M.Z.; Xu, N.W.; Zhang, C.P. Fractal analysis of acoustic emission during uniaxial and Triaxial loading of rock. *Int. J. Rock Mech. Min. Sci.* **2015**, *79*, 1365–1609. [CrossRef]

41. Pei, J.L. Study on Mechanical Behavior of Natural Multi-Group Fractured Rocks. Doctor Dissertation, Sichuan University, Chengdu, China, 2011.

Article

Experimental Study on the Characteristics of Activated Coal Gangue and Coal Gangue-Based Geopolymer

Weiqing Zhang [1,2], Chaowei Dong [1,2], Peng Huang [2,3,*], Qiang Sun [2,3], Meng Li [1,2] and Jun Chai [1,2]

[1] State Key Laboratory of Coal Resources and Safe Mining, China University of Mining and Technology, Xuzhou 221116, China; wq.zhang@cumt.edu.cn (W.Z.); dongchaowei@cumt.edu.cn (C.D.); limeng1989@cumt.edu.cn (M.L.); TS19020003A31@cumt.edu.cn (J.C.)

[2] School of Mines, China University of Mining and Technology, Xuzhou 221116, China; 5980@cumt.edu.cn

[3] Key Laboratory of the Ministry of Education for Deep Coal Mining, China University of Mining and Technology, Xuzhou 221116, China

[*] Correspondence: TBH162@cumt.edu.cn

Received: 14 April 2020; Accepted: 14 May 2020; Published: 15 May 2020

Abstract: Coal gangue-based geopolymer (CGGP) is one of the hot spots existing in the recycling of coal gangue resources due to its good comprehensive mechanical properties. However, the coal gangue structure is stable and reactivity is poor, so the coal gangue needs to be activated before utilization. In this paper, the microstructure changes of activated coal gangue by different mechanical and thermal activation methods, as well as the mechanical properties and microstructure changes of the CGGP specimens were studied by experimental investigation. The results indicated that mechanical activation and thermal activation were two effective methods to change the reactivity of coal gangue, which consisted of destroying the stable kaolinite structure and improving the activity of coal gangue. Conversely, part of the amorphous structure in coal gangue was destroyed when the activation temperature reached 900 °C, which was not conducive to the further enhancement of coal gangue activity. For the CGGP prepared by thermally activated coal gangue and modified sodium silicate alkali solution, the uniaxial compressive strength of the CGGP specimens decreased with thermal activation temperatures of the raw coal gangue materials at 700 °C, 800 °C, and 900 °C. The main reason for this was the lower amount of the active metakaolin structure in coal gangue at 900 °C, which was not conducive to the geopolymerization process.

Keywords: coal gangue; geopolymer; mechanical activation; thermal activation; mechanical property; microstructure

---

1. Introduction

China is one of the large energy producer and consumer countries in the world, mainly relying on coal. Long-term and large-scale development of coal resources lead to massive accumulation of coal gangue [1,2]. In a coal mining environment, the large amount of coal gangue piled up on the ground encroaches the land, and pollutes both the air and water environments [3–5], seriously threatening the ecological security of the mining area and the sustainable development of the coal mining industry. Therefore, resource utilization of coal gangue is urgently required. Coal gangue are mainly reused as supplementary materials in backfilling technology, building construction, energy fuel, soil improvement, water purification, special cementitious materials, etc. [6–8]. Geopolymer is a kind of environmentally friendly mineral polymer material, which has an excellent cementitious property and is a hot topic in the field of industrial solid waste recycling. Geopolymer is an inorganic polymer mainly produced by the polycondensation of aluminosilicate mineral materials with alkaline solution at room temperature, generally below 100 °C [9,10]. Geopolymer has the characteristics of high strength, fast hardness, high toughness, acid and alkali resistance, high temperature resistance, and other excellent properties. At present, geopolymers are widely used in the fields of green building materials, fire-resistant and high temperature resistant materials, sealing of toxic and dangerous wastes, aerospace materials, etc. [11,12]. The raw materials for preparing geopolymer are mainly composed of silica and alumina, including the natural clay mineral kaolinite, metakaolinite [13,14] and various industrial solid wastes, such as slag, fly ash, red mud, waste glass, and coal gangue [15–18]. The silica and alumina components account for 60–95% of coal gangue so that coal gangue has potential characteristics as a raw material to prepare geopolymer.

However, the coal gangue structure is stable. Its reactivity needs to be improved before use by activation methods, such as mechanically grinding, thermal and chemical activation [19,20]. Different activation methods have different activation effects because of the diversity and complexity of coal gangue. The activated coal gangue can be used alone or combined with other industrial solid wastes for the geopolymer reaction. The calcined coal gangue was mixed with alkaline solutions to prepare amorphous alkali-aluminosilicate cementitious materials, in which the cementitious materials prepared by modified water glass had higher compressive strength [21]. The impacts of sodium hydroxide modulus, alkali lye amount, and liquid–solid ratio on the strength and microstructure of coal gangue geopolymer were explored by Yi, et al [22]. With the increase of sodium hydroxide solution concentration, the strength of geopolymer materials increased, while the mass ratios and liquid–solid ratios did not increase linearly. The spontaneous coal gangue mixed with slag, fly ash, and alkaline solutions was used to prepare the spontaneous combustion gangue-slag-fly ash geopolymer [23,24]. The geopolymer's performance met the standard of Portland cement, and the modulus of water glass; the amount and type of activator can influence its mechanical strength. For coal gangue polymer mortars, the addition of slag and slaked lime improved the compressive strength of geopolymer [25]. The geopolymer recycled concrete was produced by hypergolic or calcined coal gangue, slag, fly ash, and recycled aggregate as raw materials. Its compressive and splitting tensile strengths were high and in line with the engineering requirements of concrete [12]. Other geopolymers with excellent properties were also prepared by coal gangue and thermal activated sludge [26,27] or red mud [28,29]. After adding fly ash, the geopolymer prepared by coal gangue, red mud, and fly ash with sodium silicate solution can reach a compressive strength of 7.3 MPa [30]. In addition, coal gangue can be used as aggregate to be cemented or compounded by other geopolymer materials. A paste filling material was prepared by cementing spontaneous combustion coal gangue with geopolymer, which was prepared by fly ash, cement, slag, and modified sodium silicate [31–33].

As a solid waste resource, coal gangue has potential value in geopolymer research. Recently, most of the researches have focused on experimental research of different material ratios, but the role of each component in the geopolymer preparation process was not completely revealed. This is of great significance for the theoretical guidance of geopolymer preparation using coal gangue. Besides, the activation effect changes with the diversity and complexity of coal gangue. Therefore, this paper mainly studied the microstructure change of coal gangue under different mechanical and thermal activation conditions. The active structural characteristics of activated coal gangue as raw material for preparing geopolymer were summarized. The macro mechanical property and micro structural characteristics of coal gangue-based geopolymer (CGGP) were measured and analyzed for in depth understanding of the polymerization of CGGP.

2. Materials and Methods

2.1. Coal Gangue Sample

Raw coal gangue samples were obtained after washing raw coal from a coal mine in mid-eastern China. The raw coal gangue was broken and screened to 80–100 mesh and dried for 6 h at 30 °C air environment to obtain the experimental sample, which was named as-received coal gangue, as shown in Figure 1.

(a)

(b)

Figure 1. Coal gangue samples. (**a**) Raw coal gangue; (**b**) As-received coal gangue.

The chemical composition of as-received coal gangue was measured by Bruker S8 Tiger X-Ray Fluorite Spectroscopy (XRF) at a voltage of 50 kV and current of 50 mA with the no-standard quantitative analysis. The loss on ignition was tested according to the international standard ASTM D7348-13. The main chemical compositions and loss on ignition of as-received coal gangue are shown in Table 1. The proximate and elemental analysis (air-dried base) are shown in Table 2.

Table 1. Main chemical composition and loss on ignition of as-received coal gangue (wt%).

Chemical Composition								Loss on Ignition
SiO_2	Al_2O_3	Fe_2O_3	CaO	K_2O	TiO_2	MgO	Na_2O	
45.26	21.62	2.839	2.69	1.65	0.72	0.608	0.415	23.4

Table 2. Proximate and elemental analysis of as-received coal gangue (wt%).

Proximate Analysis			Elemental Analysis					
Ash Content	Volatile	Fixed Carbon	Moisture	Carbon	Oxygen	Hydrogen	Nitrogen	Sulfur
77.76	12.46	8.87	0.91	12.42	7.12	1.27	0.26	0.26

Table 1 shows that the main chemical components of as-received coal gangue were SiO_2 and Al_2O_3, accounting for 66.88%, according to the quantitative analysis results of XRF standard-free samples. Table 2 shows that the ash content of as-received coal gangue was the highest, and the proportions of volatile and fixed carbon were relatively higher as per proximate analysis proportions. On the other hand, the elemental composition analysis showed that the content of carbon element was the highest, followed by that of oxygen.

2.2. Methods and Processes

2.2.1. Activation of As-Received Coal Gangue

The as-received coal gangue was ground by a planetary mill (Instrument type: BM6Pro) continuously for 2 h, 10 h, and 20 h at 400 rpm to obtain mechanical activation coal gangue (named as 2 h-CG, 10 h-CG, and 20 h-CG respectively). Each grinding was carried out in a 250 ml stainless steel pot using 300 g zirconia balls (ball gradation was 116:2198 and weights were 0.86 g and 0.091 g respectively). The three grinding times were chosen according to the reference [19].

The as-received coal gangue after being ground for 2 h was calcined for 2 h at 700 °C, 800 °C, and 900 °C in a tube furnace (Instrument type: MTF 12/38/250/301) to obtain thermal activation coal gangue (named as 700 °C-CG, 800 °C-CG and 900 °C-CG respectively). The three calcination temperatures were determined according to the transformation temperature of the kaolinite structure to active metakaolinite [10].

2.2.2. Preparation of Alkali Medium

A certain amount of water glass (8.5% Na_2O, 26.5% SiO_2, 65% H_2O,) was first mixed with deionized water with magnetic stirring at low speed at 30 °C for 20 min, then slowly poured into a beaker containing a certain amount of 96% sodium hydroxide. The mixture was fully stirred to obtain an even alkali medium of modified sodium silicate with modulus of 1 and solid content of 30%. The alkali medium was used after it cooled to room temperature.

2.2.3. Preparation of CGGP

The preparation process of CGGP was as follows: place the thermally activated coal gangue into a mixing pot, add the alkali medium (mass ratio of liquid to solid was 11:14), and mechanically stir the solid–liquid mixture until it becomes uniform; by this means the CGGP sample was obtained. The sample was filled into a mold with diameter of 50 mm and height of 100 mm, cured at room temperature until it solidified, then removed from the mold and cured continuously for 7 days at 20 °C and 92% relative humidity to obtain the CGGP specimen. The preparation process is shown in Figure 2. The CGGP specimens were named as 700 °C-CGGP, 800 °C-CGGP and 900 °C-CGGP, respectively. The three CGGP specimens, prepared mainly by thermally activated coal gangue and medium alkali, were determined based on the prior knowledge of the importance of the thermal temperature on activation results [10,19] according to sensitivity analysis [34–36].

Figure 2. Preparation process of coal gangue-based geopolymer (CGGP) specimen: (**a**) 700 °C-CG; (**b**) Alkali medium; (**c**) Mixing and stirring; (**d**) Filling into mold; (**e**) Solidifying at room temperature; (**f**) Cured 700 °C-CGGP specimen.

2.3. Sample Characterization

The mineral composition was measured by Bruker D8 advance X-Ray Diffraction (XRD) at scanning speed 2°/min and scanned area 5–70° (2θ). Jade 6.5 software was used for interpretation of XRD-patterns. The functional groups were measured by Bruker Vertex 80 v Fourier Transform Infrared Spectroscopy (FTIR) with spectral range of 400–4000 cm^{-1} and resolution of 0.06 cm^{-1}. Omnic software was used for interpretation of FTIR spectra.

The uniaxial compressive strength (UCS) values were tested according to the American Society for Testing Materials (ASTM) by an electro-hydraulic servo universal test machine (Instrument type: WAW-1000D, Changchun Sinter Testing Machine Co., Ltd., China). The displacement rate was 0.05 mm/min. The test for each type of CGGP sample was repeated three times, and its average value was taken as the final UCS.

3. Results and Discussion

3.1. Effect of Mechanical Activation on Coal Gangue Structure

The structure of the silicon–alumina phase in coal gangue, the main raw materials for preparing geopolymer, mainly exists in the form of a stable kaolinite structure. The destruction of the kaolinite crystal structure is an important aspect for coal gangue activation [37]. The XRD spectra of as-received coal gangue and mechanical activation coal gangue are shown in Figure 3.

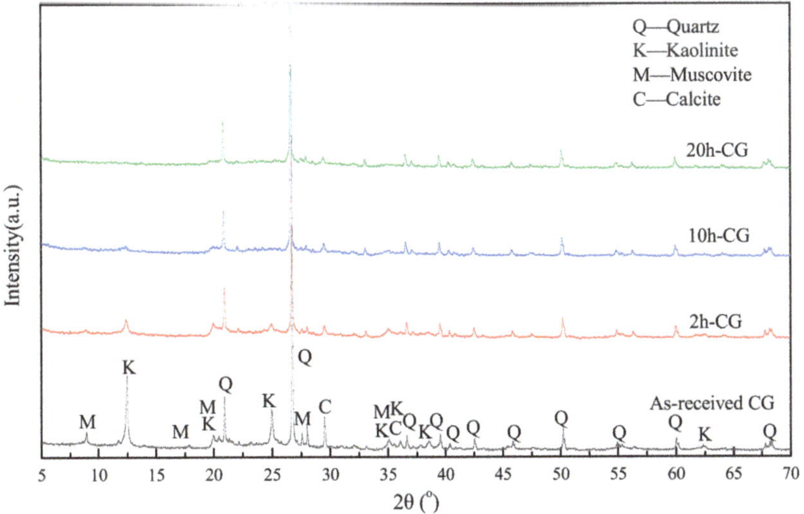

Figure 3. XRD spectra of as-received coal gangue and mechanical activation coal gangue.

Figure 3 shows that the as-received coal gangue was mainly composed of minerals such as kaolinite, quartz, muscovite, and calcite. According to the XRD spectra of raw coal gangue and mechanically activated coal gangue: (1) The intensity of the diffraction peaks at 12.46°, 24.99°, 35.11°, and 38.55° associated with the kaolinite structure decreased significantly after grinding for 2 h, became very weak after grinding for 10 h, and disappeared completely after grinding for 20 h. The intensity reduction of the kaolinite structure indicated that the stable crystal structure of kaolinite in coal gangue can be effectively destroyed by mechanical grinding. (2) The intensity of the diffraction peaks of muscovite at 8.94° and calcite at 29.5° weakened at 2 h grinding and disappeared at 10 h grinding, indicating that the structures of muscovite and calcite were also destroyed. (3) The diffraction peaks of quartz at 20.96° and 26.73° were stable during the mechanical grinding process and the peak intensities were always significant, which showed that the quartz structure in coal gangue is stable.

The XRD results demonstrated that the mineral crystal structure in coal gangue had been destroyed to different degrees except for quartz under the condition of mechanical activation. An obvious change was observed with the destroyed kaolinite structure, which showed that mechanical activation was conducive to improve the activity of coal gangue.

The structural unit layer of kaolinite in coal gangue is a 1:1 stacked layer composed of Si^{IV}–O tetrahedral layer and Al^{VI}–O octahedral layer, and the stable structural unit layers are linked by hydrogen bonding [38]. Therefore, the removal of the hydroxyl structure in coal gangue can reduce the effect of hydrogen bonding, thus effectively destroying the order of the kaolinite structure and enhancing the activity of coal gangue [39]. The change of the hydroxyl groups can be effectively characterized by FTIR spectra. The FTIR spectra of as-received coal gangue and mechanically activated coal gangue are shown in Figure 4.

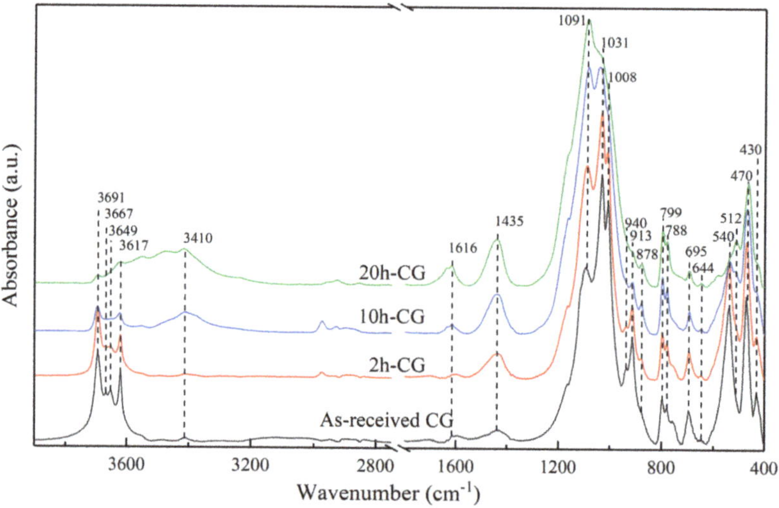

Figure 4. FTIR spectra of as-received coal gangue and mechanically activated coal gangue.

Figure 4 displays that with the increase of mechanical activation time, the peaks in the range of 3800–3200 cm⁻¹, 1200–950 cm⁻¹ and 650–400 cm⁻¹ changed significantly. According to the FTIR spectra: (1) In the range of 3800–3600 cm⁻¹, the peaks at 3691 cm⁻¹, 3667 cm⁻¹, and 3649 cm⁻¹ belong to the inner surface hydroxyl stretching vibration, and the peaks at 3617 cm⁻¹ belong to the inner hydroxyl stretching vibration of Al^{VI}–O octahedron [37,40,41]. The broad peak in the range of 3600–3200 cm⁻¹ is related to hydrogen bonding of water molecules. With the increase of the mechanical grinding time, the peaks of hydroxyl stretching vibration in the range of 3800–3600 cm⁻¹ gradually weakened, while the broad peaks at 3410 cm⁻¹ began to increase significantly from 10 h, indicating that with the removal of hydroxyl in the kaolinite structure, adsorbed water gradually formed. After 20 h grinding, only weak absorption peaks of hydroxyl stretching vibration could be observed, while the vibration peaks of hydrogen bonding between water molecules was further enhanced. This showed that the hydroxyl structure in mechanically activated coal gangue decreases and the amount of adsorbed water increases with the increase in grinding time. (2) The peaks at 940 cm⁻¹ and 913 cm⁻¹ belong to the bending vibration of inner surface hydroxyl and inner hydroxyl, respectively [40,42]. With the increase in mechanical grinding time, the two peaks gradually weakened and disappeared at 20 h, which further confirmed the removal of the hydroxyl structure in mechanically activated coal gangue. At the same time, the changing pattern of the bending vibration peak [22,37] at 1616 cm⁻¹ was consistent with that at 3410 cm^{-1}, which also confirmed the gradual formation of adsorbed water. (3) The Si–O stretching vibration peaks in the range of 1200–950 cm^{-1} were mainly at 1091 cm^{-1}, 1031 cm^{-1}, and 1008 cm^{-1}, which belonged to the different stretching vibration of Si–O–Si in the Si^{IV}–O tetrahedron [43]. With the increase in grinding time, the peaks at 1031 cm^{-1} and 1008 cm^{-1} gradually weakened and disappeared, and the peaks at 1091 cm^{-1} increased significantly, which indicated that the main structure related to Si–O in coal gangue had changed, mainly with the Si^{IV}–O tetrahedral structure partly deformed. Such deformation was mainly caused by the change of bond length and bond angle of the Si^{IV}–O tetrahedron which was affected by the change of the Al–O polyhedral structure. At 20 h, the peak center shifted from 1031 cm^{-1} to 1091 cm^{-1}, indicating the formation of amorphous structure in mechanically activated coal gangue [19]. (4) The peak at 878 cm^{-1} belongs to the tetrahedral structure of Al^{IV}–O [22]. With the increase in grinding time, the peak intensity underwent a gradual increase, indicating the rise of the tetrahedral structure of Al^{IV}–O, which was beneficial to the enhancement of the activity of coal gangue. (5) The peaks at 540 cm^{-1}, 512 cm^{-1} and 430 cm^{-1} are mainly attributed to the bending vibration of Si–O–Al^{VI}, also the contribution from the deformation vibration of Si–O–Si [43]. The peak intensity at 540 cm^{-1} and 430 cm^{-1} decreased with the increase in grinding time and disappeared at 20 h, indicating that the structure of Si–O–Al^{VI} in mechanically activated coal gangue declined. This was caused by the removal of hydroxyl groups. The peak intensity at 512 cm^{-1} increased with the increase in grinding time, enhancing the formation of a new Si–O–Al structure. (6) The peak at 1435 cm^{-1} was attributed to the O–C–O stretching vibration in CO$_3^{2-}$, which means that coal gangue may contain carbonates [25,44]. With the increase in grinding time, the peak of the spectrum increased, which is believed to be caused by carbonation during the sample grinding process [22]. (7) The deformation vibration of Si–O chains in the range of 800–650 cm^{-1} [38] and the O–Si–O vibration of quartz structure at 470 cm^{-1} had no obvious changes after the mechanical grinding process.

With the increase in mechanical activation time, the inner surface hydroxyl and inner hydroxyl in coal gangue were gradually removed. The Si–O structure changed, the Si–O–Al^{VI} structure decreased and the Al^{IV}–O tetrahedron structure increased. Such changes were conducive to improving the activity of coal gangue.

3.2. Effect of Thermal Activation on the Coal Gangue Structure

The fixed carbon content in coal gangue is not conducive to the enhancement of the activity of coal gangue; this is why the thermal activation method was considered to be more useful to effectively remove the fixed carbon. The XRD spectra of thermally activated coal gangue (named as uncalcined-CG) and non-thermally activated coal gangue are shown in Figure 5.

From Figure 5: (1) The diffraction peaks of the kaolinite crystal structure at 12.46° and 24.99° disappeared completely in the thermally activated coal gangue, indicating that the thermal activation had a significant destructive effect on the kaolinite structure of coal gangue. (2) The diffraction peak of calcite at 29.51° disappeared completely in the thermally activated coal gangue, and this means that thermal activation can effectively destroy the calcite structure in coal gangue. (3) The diffraction peaks of quartz at 20.96° and 26.73° were significant throughout the process of thermal activation, indicating the quartz structure in coal gangue was stable. (4) In the coal gangue thermally activated at 700 °C and 800 °C, new diffraction peaks at 25.58° and 31.46° appeared, but the intensity weakened obviously at 900 °C, indicating that there were new structures in thermally activated coal gangues but these structures were destroyed at higher temperature.

As observed, under the condition of thermal activation, the mineral components in coal gangue except the quartz structure were significantly damaged, which was conducive to the activity change of coal gangue. However, if the thermal activation temperature was too high, part of the new structures in coal gangue were also destroyed, which is not useful for further improvement of coal gangue activity.

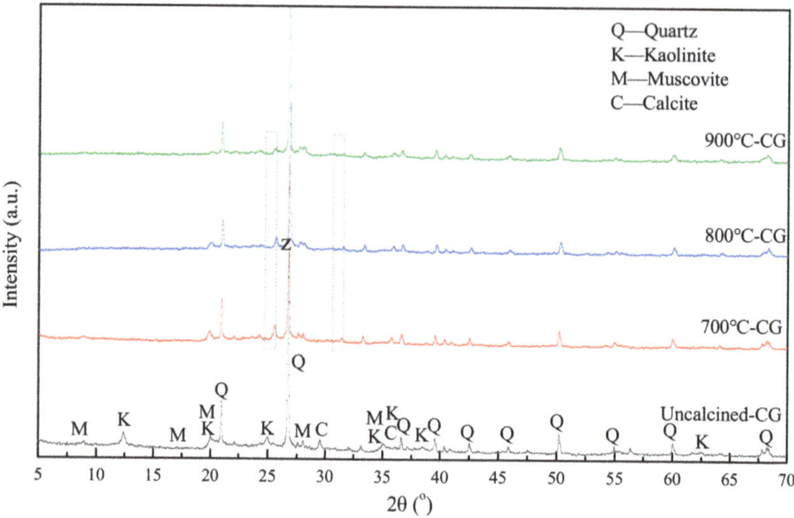

Figure 5. XRD spectra of thermally activated coal gangue and non-thermally activated coal gangue.

The FTIR spectra of thermally activated coal gangue and non-thermally activated coal gangue (also named as uncalcined-CG) are shown in Figure 6.

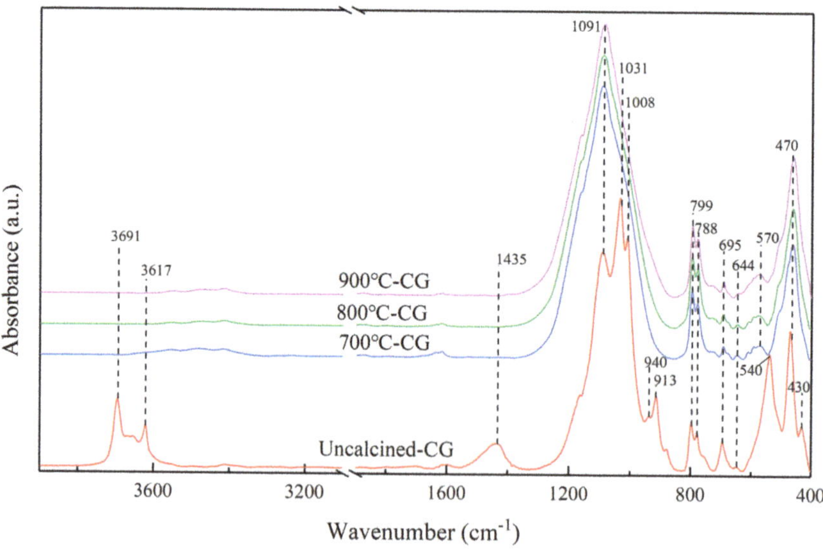

Figure 6. FTIR spectra of thermally activated coal gangue and uncalcined-CG.

From Figure 6: (1) The peak of hydroxyl stretching vibration in the range of 3800–3600 cm^{-1} and the peaks of hydroxyl bending vibration at 940 cm^{-1} and 913 cm^{-1} of thermally activated coal gangue disappeared completely, indicating that the hydroxyl structure in coal gangue can be effectively removed by thermal activation. (2) The stretching vibration peaks of Si–O in the range of 1200–950 cm^{-1} changed significantly. Compared with the uncalcined-CG, the peaks at 1031 cm^{-1} and 1008 cm^{-1} disappeared, while the peak at 1091 cm^{-1} increased and widened significantly, indicating that the main structures related to Si–O in coal gangue changed significantly, especially the crystal kaolinite structure which changed into the amorphous metakaolinite structure [19]. Such structure changes were similar to those in mechanically activated coal gangue for 20 h grinding time. So, the desired results by thermal activation can be achieved by a sufficiently long mechanical activation process. (3) The peaks of Si–O–Al and O–Si–O in the range of 650–400 cm^{-1} changed significantly. The peaks of Si–O–Al^{VI} at 540 cm^{-1} and 430 cm^{-1} disappeared in the thermally activated coal gangue, and the new peaks of Si–O–Al^{IV} at 570 cm^{-1} appeared. Such results indicated a significant decrease of Al^{VI}–O structure in the kaolinite structure and the formation of a new Al^{VI}–O structure in the metakaolinite structure [19,45]. (4) The complete disappearance of CO_3^{2-} structure at 1435 cm^{-1} after thermal activation indicated the destruction of the carbonate structure in coal gangue, which was consistent with the disappearance of the calcite peak in XRD (Figure 5). (5) There were no obvious changes to the peaks in the range of 800–650 cm^{-1} and 470 cm^{-1}.

Briefly, under the condition of thermal activation, the hydroxyl groups in coal gangue were completely removed and the kaolinite structure was significantly destroyed. The appearance of the Al^{IV}–O structure representing the amorphous metakaolinite structure indicated the improvement of coal gangue activity, so, thermally activated coal gangue can be used as raw material for geopolymer preparation.

3.3. Analysis of Macro- and Micro-Properties of CGGP

The CGGP, with excellent mechanical properties, can be prepared by the geopolymerization of activated coal gangue and alkali medium [10,22]. The UCS is one of the main and most widely used testing methods for the mechanical properties of rocks in mining engineering projects [46,47]. The UCS results of the three CGGP specimens are shown in Figure 7.

Figure 7. Uniaxial compressive strength (UCS) results of the CGGP specimens.

Figure 7 shows that the average UCS results of the three CGGP specimens fluctuated in the range of 8.55–17.85 MPa, which gives higher UCS results than the previous gangue-cemented paste backfill [48,49]. With the increase in the thermal activation temperature of coal gangue materials, the UCS of CGGP specimens increased first and then decreased. The CGGP specimen prepared by the thermal activation at 800 °C had the largest average UCS, 17.85 MPa, which was 66% (+/−17%) and 110% (+/−21%) higher than those recorded at 700 °C-CGGP and 900 °C-CGGP, respectively.

The main reason was that the presence of more active metakaolinite structures in the thermal activation coal gangue at 800 °C (according to Figures 5 and 6) gave stronger coal gangue reactivity and this became more beneficial to the geopolymerization process. When the temperature increased to 900 °C, part of the active structure in coal gangue was destroyed, resulting in a decrease in coal gangue reactivity [50], which was not conducive to the geopolymerization; thus, the compressive strengths of CGGP specimens were relatively weak. The XRD spectra of CGGP specimens are shown in Figure 8.

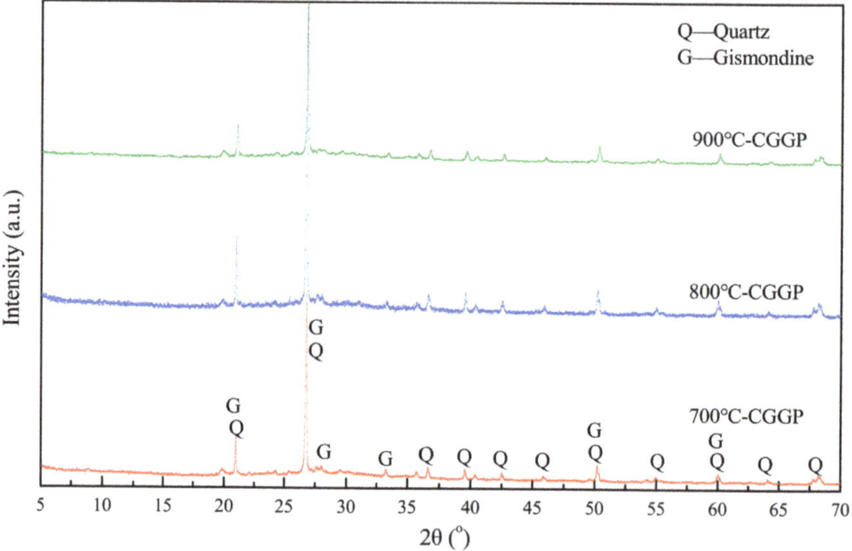

Figure 8. XRD spectra of CGGP specimens.

Figure 8 shows that the crystalline structure of the CGGP specimens was simple, especially the quartz structure. The new diffraction peaks at 25.58° and 31.46° (Figure 5) disappeared in CGGP specimens, indicating that these structures had a great impact in the geopolymerization. The FTIR spectra of CGGP specimens are shown in Figure 9.

From Figure 9: (1) The main peaks of the three CGGP specimens were almost the same, indicating the similarity of the main structures of CGGP. However, compared with the thermally activated coal gangue (700 °C-CG), there were significant changes mainly in the ranges of 3800–3200 cm^{-1} and 1200–950 cm^{-1}. (2) In CGGP, the broad peak of hydroxyl stretching vibration at 3440 cm^{-1} and the bending vibration peak of hydroxyl at 1653 cm^{-1} characterized the chemical binding water generated during geopolymerization [22]. The center of the Si–O stretching vibration spectrum in the range of 1200–950 cm^{-1} shifted from 1085 cm^{-1} to 1011 cm^{-1} and was extended, which was an important point, characterizing the existence of a large number of Si–O–Si and Si–O–Al^{IV} structures in the geopolymer [22,25,30]. The appearance of the Al^{IV}-O tetrahedral structure peak at 878 cm^{-1} confirmed such a conclusion. (3) The difference between the three CGGP peaks was mainly the variation of the peak intensity in the 1200–950 cm^{-1} range. The peak intensity of CGGP prepared by thermal activation of coal gangue at 900 °C was relatively weak at 1011 cm^{-1}, which indicated that the structure of Si–O–Si and Si–O–Al^{IV} in the CGGP structure was less, and the geopolymerization degrees of the 700 °C-CGGP and 800 °C-CGGP were weaker. The peaks of the Al^{IV}–O tetrahedral structure at 878 cm^{-1} were also weakened, which confirmed the same conclusion. So, the new structure in the 900 °C-CGGP was relatively less, resulting in a relatively lower UCS result of the 900 °C-CGGP specimen.

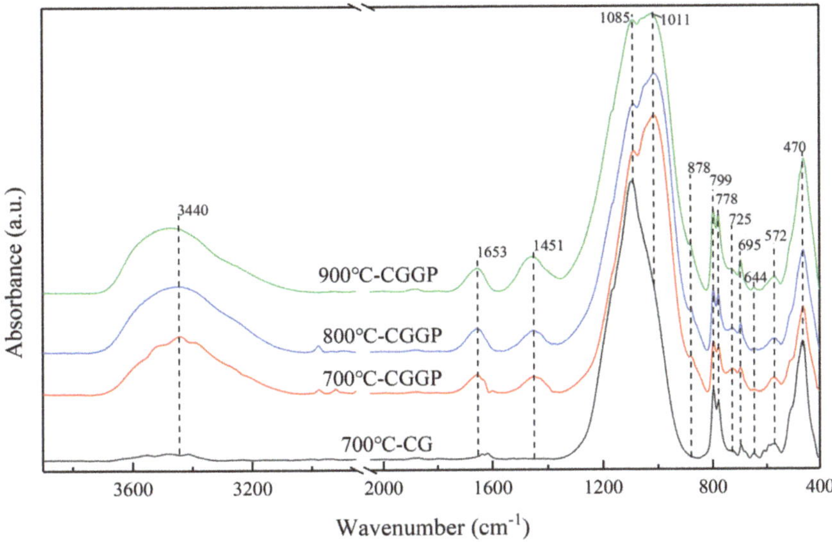

Figure 9. FTIR spectra of CGGP specimens.

In sum, the microstructure of CGGP was significantly different from that of the thermally activated coal gangue, indicating the formation of a new structure in CGGP. The relatively lower content of the new structure was the main reason for the lower UCS result of the CGGP.

4. Conclusions and Future Prospects

In this paper, the microstructure changes of coal gangue activated by different mechanical and thermal activation methods, as well as the macro-mechanical properties and microstructure changes of the CGGP were studied. The main conclusions are as follows:

(1) Mechanical activation and thermal activation are two effective methods for changing the reactivity of coal gangue, whereby both can destroy the stable kaolinite structure and improve the activity of coal gangue. Both FTIR and XRD spectra showed the removal of inner surface hydroxyl and inner hydroxyl, a decrease of Si–O–Al^{VI} structure, and an increase of Al^{IV}–O structure in the activated coal gangue.

(2) The activation effect and microstructure change of coal gangue were different under different activation methods. With the increase in mechanical activation time, the crystal structure of kaolinite in coal gangue decreased and disappeared completely after 20 h grinding. Thermal activation can completely remove the hydroxyl in the kaolinite structure and form the active metakaolinite structure in coal gangue. However, when the activation temperature was too high (900 °C), the new structure in thermally activated coal gangue was destroyed, which was disadvantageous to the further improvement of the coal gangue activity.

(3) The structure of activated coal gangue can be destroyed by the alkaline medium and reconstituted to form CGGP. The UCS results of CGGP prepared by different thermal activations of coal gangue were different. The UCS result of 800 °C-CGGP was the largest, which was higher than those recorded with 700 °C-CGGP and 900 °C-CGGP, respectively. In this paper, results showed that the optimized thermal activation temperature for coal gangue is about 800 °C.

(4) The reason for the lower compressive strength of CGGP was that the active metakaolinite structures were destroyed in thermally activated coal gangue, which led to the decrease of the coal gangue activity and appeared to be a disadvantage for the geopolymerization. FTIR and

XRD spectra showed the formation of new structures in CGGP, while the new structures in CGGP with low compressive strengths were relatively lower.

Coal gangue is a good inorganic material with an aluminosilicate structure for the preparation of geopolymers. Further research work needs to be carried out around the following respects: (1) quantitative analysis of the new structure composition in CGGP to reveal the reaction mechanism; (2) collecting more kinds of coal gangue samples in order to study their activation characteristics using different activation methods; and (3) research on the application of CGGP at the mining engineering site, such as backfilling materials in gob or filling body besides the roadway along the gob in coal mines.

Author Contributions: All of the authors contributed extensively to the present paper. W.Z. and P.H. conceived and provided theoretical and methodological guidance in the research. W.Z. and C.D. designed the experiments, processed and analyzed the data. C.D. and J.C. performed the experiments. M.L., Q.S., and P.H. reviewed and revised the manuscript extensively. All authors have read and agreed to the published version of the manuscript.

Funding: This research was funded by the National Key R&D Program of China (grant number 2018YFC0604704), the Independent Research Project of State Key Laboratory of Coal Resources and Safe Mining, CUMT (SKLCRSM19X001) and the National Natural Science Foundation of China (51704282).

Acknowledgments: We thank the anonymous reviewers for constructive and enlightened comments and suggestions in the revising process.

Conflicts of Interest: The authors declare no conflict of interest.

References

1. Querol, X.; Zhuang, X.; Font, O.; Izquierdo, M.; Alastuey, A.; Castro, I.; Van Drooge, B.L.; Moreno, T.; Grimalt, J.O.; Elvira, J.; et al. Influence of soil cover on reducing the environmental impact of spontaneous coal combustion in coal waste gobs: A review and new experimental data. *Int. J. Coal Geol.* **2011**, *85*, 2–22. [CrossRef]
2. Song, L.; Liu, S.; Li, W. Quantitative Inversion of Fixed Carbon Content in Coal Gangue by Thermal Infrared Spectral Data. *Energies* **2019**, *12*, 1659. [CrossRef]
3. Zhang, J.X.; Ju, Y.; Zhang, Q.; Ju, F.; Xiao, X.; Zhang, W.Q.; Zhou, N.; Li, M. Low ecological environment damage technology and method in coal mine. *J. Min. Strata Contr. Eng.* **2019**, *1*, 1–13.
4. Huang, Y.-L.; Zhang, J.; Yin, W.; Sun, Q. Analysis of Overlying Strata Movement and Behaviors in Caving and Solid Backfilling Mixed Coal Mining. *Energies* **2017**, *10*, 1057. [CrossRef]
5. Querol, X.; Izquierdo, M.; Monfort, E.; Alvarez, E.; Font, O.; Moreno, T.; Alastuey, A.; Zhuang, X.; Lu, W.; Wang, Y. Environmental characterization of burnt coal gangue banks at Yangquan, Shanxi Province, China. *Int. J. Coal Geol.* **2008**, *75*, 93–104. [CrossRef]
6. Li, J.; Wang, J. Comprehensive utilization and environmental risks of coal gangue: A review. *J. Clean. Prod.* **2019**, *239*, 117946. [CrossRef]
7. Zhang, Q.; Zhang, J.; Wu, Z.; Chen, Y. Overview of Solid Backfilling Technology Based on Coal-Waste Underground Separation in China. *Sustainability* **2019**, *11*, 2118. [CrossRef]
8. Huang, P.; Spearing, A.J.S.; Ju, F.; Jessu, K.; Wang, Z.; Ning, P. Control Effects of Five Common Solid Waste Backfilling Materials on In Situ Strata of Gob. *Energies* **2019**, *12*, 154. [CrossRef]
9. Davidovits, J. Geopolymers. *J. Therm. Anal. Calorim.* **1991**, *37*, 1633–1656. [CrossRef]
10. Davidovits, J. Geopolymers Ceramic-Like Inorganic Polymers. *J. Ceram. Sci. Technol.* **2017**, *8*, 335–350.
11. McLellan, B.C.; Williams, R.; Lay, J.; Van Riessen, A.; Corder, G. Costs and carbon emissions for geopolymer pastes in comparison to ordinary portland cement. *J. Clean. Prod.* **2011**, *19*, 1080–1090. [CrossRef]
12. Liu, C.; Deng, X.; Liu, J.; Hui, D. Mechanical properties and microstructures of hypergolic and calcined coal gangue based geopolymer recycled concrete. *Constr. Build. Mater.* **2019**, *221*, 691–708. [CrossRef]
13. Hajjaji, W.; Andrejkovicova, S.; Zanelli, C.; Alshaaer, M.; Dondi, M.; Labrincha, J.; Rocha, F. Composition and technological properties of geopolymers based on metakaolin and red mud. *Mater. Des.* **2013**, *52*, 648–654. [CrossRef]
14. Zhang, H.; Kodur, V.; Wu, B.; Cao, L.; Qi, S.L. Comparative Thermal and Mechanical Performance of Geopolymers derived from Metakaolin and Fly Ash. *J. Mater. Civ. Eng.* **2016**, *28*, 04015092. [CrossRef]

15. Part, W.K.; Ramli, M.; Cheah, C.B. An overview on the influence of various factors on the properties of geopolymer concrete derived from industrial by-products. *Constr. Build. Mater.* **2015**, *77*, 370–395. [CrossRef]

16. He, J.; Zhang, J.; Yu, Y.; Zhang, G. The strength and microstructure of two geopolymers derived from metakaolin and red mud-fly ash admixture: A comparative study. *Constr. Build. Mater.* **2012**, *30*, 80–91. [CrossRef]

17. Zhuang, X.Y.; Chen, L.; Komarneni, S.; Zhou, C.; Tong, D.S.; Yang, H.M.; Yu, W.H.; Wang, H. Fly ash-based geopolymer: Clean production, properties and applications. *J. Clean. Prod.* **2016**, *125*, 253–267. [CrossRef]

18. Mohamad, J.M.; Rassoul, A.; Alborz, H. Preparation and application of alkali-activated materials based on waste glass and coal gangue: A review. *Constr. Build. Mater.* **2019**, *221*, 84–98.

19. Guo, Y.; Yan, K.; Cui, L.; Cheng, F. Improved extraction of alumina from coal gangue by surface mechanically grinding modification. *Powder Technol.* **2016**, *302*, 33–41. [CrossRef]

20. Li, Y.; Yao, Y.; Liu, X.; Sun, H.; Ni, W. Improvement on pozzolanic reactivity of coal gangue by integrated thermal and chemical activation. *Fuel* **2013**, *109*, 527–533. [CrossRef]

21. Zhang, C.S.; Fang, L.M. Hardening mechanisms of alkali activated burned gangue cementitious material. *Mater. Sci. Technol.* **2004**, *12*, 597–601.

22. Cheng, Y.; Ma, H.; Hongyu, C.; Jiaxin, W.; Jing, S.; Zonghui, L.; Mingkai, Y. Preparation and characterization of coal gangue geopolymers. *Constr. Build. Mater.* **2018**, *187*, 318–326. [CrossRef]

23. Zhou, M.; Xu, M.; Li, Z.W.; Sun, Q.W. Preparation and basic properties of spontaneous combustion gangue-slag-fly ash gepolyer. *Bull. Chin. Ceram. Soc.* **2013**, *32*, 1826–1831.

24. Zhou, M.; Wang, C.Z.; Li, Z.W.; Chang, J.; Liu, Q.; Zu, X.J. Ligand optimization of spontaneous combustion coal gangue geopolymer based on the orthogonal and response surface design. *Bull. Chin. Ceram. Soc.* **2013**, *32*, 1258–1263, 1268.

25. Huang, G.; Ji, Y.; Li, J.; Hou, Z.; Dong, Z. Improving strength of calcinated coal gangue geopolymer mortars via increasing calcium content. *Constr. Build. Mater.* **2018**, *166*, 760–768. [CrossRef]

26. Xu, Z.; Zou, X.; Yang, Z. Preparation and Properties of Sludge and Coal Gangue Composite Polymer. *Asian J. Chem.* **2014**, *26*, 1751–1753. [CrossRef]

27. Xu, Z.F.; Zou, X.T.; Chen, J. Preparation of thermal activation sludge and coal gangue polymer. *Integ. Ferroelectr.* **2015**, *160*, 1–9.

28. Geng, J.; Zhou, M.; Li, Y.; Chen, Y.; Han, Y.; Wan, S.; Zhou, X.; Hou, H. Comparison of red mud and coal gangue blended geopolymers synthesized through thermal activation and mechanical grinding preactivation. *Constr. Build. Mater.* **2017**, *153*, 185–192. [CrossRef]

29. Geng, J.; Zhou, M.; Zhang, T.; Wang, W.; Wang, T.; Zhou, X.; Wang, X.; Hou, H. Preparation of blended geopolymer from red mud and coal gangue with mechanical co-grinding preactivation. *Mater. Struct.* **2016**, *50*, 109. [CrossRef]

30. Koshy, N.; Dondrob, K.; Hu, L.; Wen, Q.; Meegoda, J.N. Synthesis and characterization of geopolymers derived from coal gangue, fly ash and red mud. *Constr. Build. Mater.* **2019**, *206*, 287–296. [CrossRef]

31. Sun, Q.; Tian, S.; Sun, Q.; Li, B.; Cai, C.; Xia, Y.; Wei, X.; Mu, Q. Preparation and microstructure of fly ash geopolymer paste backfill material. *J. Clean. Prod.* **2019**, *225*, 376–390. [CrossRef]

32. Sun, Q.; Li, B.; Tian, S.; Cai, C.; Xia, Y. Creep properties of geopolymer cemented coal gangue-fly ash backfill under dynamic disturbance. *Constr. Build. Mater.* **2018**, *191*, 644–654. [CrossRef]

33. Sun, Q.; Cai, C.; Zhang, S.; Tian, S.; Li, B.; Xia, Y.; Sun, Q. Study of localized deformation in geopolymer cemented coal gangue-fly ash backfill based on the digital speckle correlation method. *Constr. Build. Mater.* **2019**, *215*, 321–331. [CrossRef]

34. Reza, A.; Seyed, A.H.; Mojtaba, S.; Abbas, A.S. Updating the neural network sediment load models using different sensitivity analysis methods: A regional application. *J. Hydroinform* **2020**, in press.

35. Andrea, S. Sensitivity Analysis for Importances Assessment. *Risk. Anal.* **2002**, *22*, 579–590.

36. Sobol, I. Sensitivity analysis for non-linear mathematical models. *Math. Model. Comput.* **1993**, *1*, 407–414.

37. Hu, P.; Yang, H. Insight into the physicochemical aspects of kaolins with different morphologies. *Appl. Clay Sci.* **2013**, *74*, 58–65. [CrossRef]

38. Frost, R.L. Hydroxy deformation in kaolin. *Clays Clay Miner.* **1998**, *46*, 280–289. [CrossRef]

39. Ptáček, P.; Frajkorová, F.; Šoukal, F.; Opravil, T. Kinetics and mechanism of three stages of thermal transformation of kaolinite to metakaolinite. *Powder Technol.* **2014**, *264*, 439–445. [CrossRef]

40. Frost, R.L. Combination Bands in the Infrared Spectroscopy of Kaolins—A Drift Spectroscopic Study. *Clays Clay Miner.* **1998**, *46*, 466–477. [CrossRef]

41. Konan, L.K.; Peyratout, C.; Smith, A.; Bonnet, J.-P.; Rossignol, S.; Oyetola, S. Comparison of surface properties between kaolin and metakaolin in concentrated lime solutions. *J. Colloid Interface Sci.* **2009**, *339*, 103–109. [CrossRef] [PubMed]

42. Hamzaoui, R.; Muslim, F.; Guessasma, S.; Bennabi, A.; Guillin, J. Structural and thermal behavior of proclay kaolinite using high energy ball milling process. *Powder Technol.* **2015**, *271*, 228–237. [CrossRef]

43. Castellano, M.; Turturro, A.; Riani, P.; Montanari, T.; Finocchio, E.; Ramis, G.; Busca, G. Bulk and surface properties of commercial kaolins. *Appl. Clay Sci.* **2010**, *48*, 446–454. [CrossRef]

44. Zhu, B.Z.; Sun, Y.L.; Xie, C.W. Spectroscopy research on the Guizhou Xingyi gangue of different calcined temperatures. *J. China Coal Soc.* **2008**, *9*, 1049–1052.

45. Vizcayno, C.; De Gutiérrez, R.M.; Castello, R.; Rodriguez, E.D.; Guerrero, C. Pozzolan obtained by mechanochemical and thermal treatments of kaolin. *Appl. Clay Sci.* **2010**, *49*, 405–413. [CrossRef]

46. Shahri, A.A.; Larsson, S.; Johansson, F. Updated relations for the uniaxial compressive strength of marlstones based on P-wave velocity and point load index test. *Innov. Infrastruct. Solut.* **2016**, *1*, 17. [CrossRef]

47. Asheghi, R.; Shahri, A.A.; Zak, M.K. Prediction of Uniaxial Compressive Strength of Different Quarried Rocks Using Metaheuristic Algorithm. *Arab. J. Sci. Eng.* **2019**, *44*, 8645–8659. [CrossRef]

48. Chen, S.; Du, Z.; Zhang, Z.; Yin, D.; Feng, F.; Ma, J. Effects of red mud additions on gangue-cemented paste backfill properties. *Powder Technol.* **2020**, *367*, 833–840. [CrossRef]

49. Jiang, H.; Fall, M.; Cui, L. Freezing behaviour of cemented paste backfill material in column experiments. *Constr. Build. Mater.* **2017**, *147*, 837–846. [CrossRef]

50. Li, H.; Sun, H.-H. Microstructure and cementitious properties of calcined clay-containing gangue. *Int. J. Miner. Met. Mater.* **2009**, *16*, 482–486. [CrossRef]

MDPI
St. Alban-Anlage 66
4052 Basel
Switzerland
Tel. +41 61 683 77 34
Fax +41 61 302 89 18
www.mdpi.com

Energies Editorial Office
E-mail: energies@mdpi.com
www.mdpi.com/journal/energies